DNA-encoded Library Technology for Drug Discovery

Drug Discovery Series

Editor-in-chief:
Simon E. Ward, *Cardiff University, UK*

Series editors:
David Fox, *Vulpine Science & Learning, UK*
Corey Hopkins, *University of Nebraska Medical Center, USA*
Ian Storer, *AstraZeneca, UK*

For a list of titles in this series see: rsc.li/drug-discovery

How to obtain future titles on publication:
A standing order plan is available for this series. A standing order will bring delivery of each new volume immediately on publication.

For further information please contact:
Book Sales Department, Royal Society of Chemistry, Thomas Graham House, Science Park, Milton Road, Cambridge, CB4 0WF, UK
Telephone: +44 (0)1223 420066, Fax: +44 (0)1223 420247
Email: booksales@rsc.org
Visit our website at books.rsc.org

DNA-encoded Library Technology for Drug Discovery

Edited by

Guansai Liu
HitGen Inc., China
Email: gs.liu@hitgen.com

Casey J. Krusemark
Purdue University, USA
Email: ckrusema@purdue.edu

and

Jin Li
HitGen Inc., China
Email: jin.li@hitgen.com

ROYAL SOCIETY OF CHEMISTRY

Drug Discovery Series No. 85

Hardback ISBN: 978-1-78801-487-8
PDF ISBN: 978-1-78801-603-2
EPUB ISBN: 978-1-83916-055-4
Print ISSN: 2041-3203
Electronic ISSN: 2041-3211

A catalogue record for this book is available from the British Library

The Royal Society of Chemistry is a charity, registered in England and Wales, Number 207890, and a company incorporated in England by Royal Charter (Registered No. RC000524), registered office: Burlington House, Piccadilly, London W1J 0BA, UK, Telephone: +44 (0)20 7437 8656.

For further information see our website at www.rsc.org

For general enquiries, please contact books@rsc.org

For EU product safety enquiries, please email books@rsc.org or contact Royal Society of Chemistry Worldwide (Germany) GmbH, Römischer Hof, Unter den Linden 10, 10117 Berlin.

Preface

After nearly 30 years of development, DNA-encoded library (DEL) technology has matured significantly and is now a widely practiced method for small molecule drug discovery. Compared to traditional small molecule screening methods, DEL technology is dramatically lower in cost and offers marked improvements in efficiency and capacity for exploring vast chemical space. Initial hits to a target protein can be identified rapidly with minimal material input. DELs have become an essential tool in drug discovery, profoundly impacting the field.

The advancement of DEL technology can be divided into three major stages. The first stage, from 1992 to 2006, was the proof-of-concept period. During this time, researchers transitioned DEL technology from theory to practice. Pioneering demonstrations by the Neri, Harbury, and Liu groups promised an exciting future where molecular biology could be harnessed for small molecule discovery. New synthetic methods suitable for on-DNA chemical synthesis were developed, and systems were designed to program chemical synthesis using DNA sequences, enabling the *in vitro* evolution of wholly synthetic molecular populations.

The second stage, from 2006 to 2012, saw DEL technology move into real-world applications in drug discovery and initial adoption by the pharmaceutical industry. This period was marked by GSK's acquisition of the DEL-based startup Praecis in 2006. The advent of next-generation DNA sequencing (NGS) during this era made the analysis of large libraries feasible for DELs prepared by DNA ligation-based tagging. This period included the landmark discovery of small molecules targeting Aurora A and p38 MAP kinase by the Praecis/GSK team. However, literature reports of DEL discoveries during this time were sparse and mostly from academic groups.

The third stage, from 2012 to the present, marks a period of rapid development and broad acceptance of DEL technology. New synthetic

Drug Discovery Series No. 85
DNA-encoded Library Technology for Drug Discovery
Edited by Guansai Liu, Casey J. Krusemark and Jin Li

Published by the Royal Society of Chemistry, www.rsc.org

methods suitable for DEL synthesis have become abundant, and DEL chemistry has matured dramatically, yielding greatly improved libraries in terms of drug-likeness. DELs have nearly become a commodity and are commercially available from several providers at reasonable prices. DEL-based discovery is now practiced by all major pharmaceutical companies, with many published examples of discovery applications for biological targets with high therapeutic interest and DEL-discovered clinical candidates. Researchers are now expanding DEL use beyond the standard, *in vitro* selection for protein binders and into new application areas. Indeed, the future looks bright for DELs in drug discovery.

Unlike prior books on DEL technology, this book does not explain the basic principles and operation methods of DEL technology. Instead, it aims to describe recent major applications and progress in the field. We invited academic and industrial experts to review and demonstrate new applications of DEL technology, including machine learning using DEL data, one-bead-one-compound DELs for traditional biochemical screens, live cell-based DEL selections, DELs with covalent warheads, molecule optimization using focused DELs, and specific DEL applications for protein degraders. We hope this book inspires readers to think about new frontiers in the application of DELs to further promote the use of DEL technology in drug discovery.

Guansai Liu
Casey J. Krusemark
Jin Li

Contents

Drug Discovery Series No. 85
DNA-encoded Library Technology for Drug Discovery
Edited by Guansai Liu, Casey J. Krusemark and Jin Li

Published by the Royal Society of Chemistry, www.rsc.org

CHAPTER 1

Affinity-based DEL Selections with Different Target Types: Overview and Achievements

QIUXIA CHEN AND DAVID I. ISRAEL*

HitGen Inc., No. 8, Huigu 1st East Road, Shuangliu District, Chengdu, China
*Email: david.israel@hitgen.com

1.1 Introduction

DNA-encoded library (DEL) technology has witnessed significant advancements since its conceptualization in 1992.[1–3] DELs are mixtures of small molecules conjugated to DNA tags, whereby small molecules are synthesized on DNA using sequential chemical reactions, each of which is coupled to ligation of an oligonucleotide tag. The DNA sequence functions as the identifier of the chemical structure of each compound in the library, thus facilitating the screening of a vast number of molecules in a mixture of library compounds. The development of combinational chemistry and next generation sequencing (NGS) has enabled the rapid investigation of DELs ranging in size from approximately a million to over a trillion compounds,[4–6] with a turnaround time as short as a month. Major pharmaceutical companies including GlaxoSmithKline (GSK), Roche, Amgen, Pfizer, Merck (MSD) and others are actively engaged in drug discovery endeavours based on DEL technology. Additionally, DEL service providers such as HitGen, WuXi AppTec and X-Chem have made this technology available to other pharmaceutical companies, smaller biotechs, and academic scientists. With the progression of this technology over the past three decades, DEL

Drug Discovery Series No. 85
DNA-encoded Library Technology for Drug Discovery
Edited by Guansai Liu, Casey J. Krusemark and Jin Li

Published by the Royal Society of Chemistry, www.rsc.org

chemistry,[7–9] the copy number of DEL molecules,[10–12] selection conditions and data analysis[13–19] have been intensively explored and improved upon. DEL selection now provides an efficient, ultra-high throughput, and cost-effective screening strategy that has made significant achievements and yielded several small molecules that have progressed into different stages of clinical development (Figure 1.1).

1.2 Affinity-based DEL Selection Strategies

Binding processes serve as the basis of all molecules that execute biological functions. DEL selection leverages this principle to investigate affinity-based binding of DEL molecules to their respective targets. Binders selected from DELs are identified through decoding of DNA sequences, and subsequent off-DNA synthesis and biological testing is conducted for hit validation. Different selection strategies have been developed over the past three decades, primarily distinguished by the methods of displaying the target and of partitioning binders from non-binders.

1.2.1 DEL Selections on a Solid Support

The majority of DEL selection processes employ a solid support such as magnetic beads or resin (*e.g.* PhyTip columns) to immobilize the target and facilitate the separation of the binders from the bulk DEL solution. The screening target is usually produced with an affinity tag such as hexa-histidine or biotin, to enable immobilization on the solid support (Figure 1.2A). The procedure encompasses several steps including target immobilization, DEL incubation with the protein or nucleic acid target, washing to eliminate non-specific or weak binders, binder release through heating, competitive ligand elution or target elution, and the utilization of the output as the input for subsequent selection rounds. The order of target immobilization and DEL incubation steps can differ in different selection modes, but they all utilize immobilization for DEL separation.[27,28] DEL selection on a solid support is primarily used due to its straightforward protocol and rapid adaptation to benchtop instruments. All the aforementioned clinical-stage drug candidates were discovered utilizing this selection strategy. Covalent-based DEL selection[29,30] operates in a similar format but binders are not released given the covalent attachment of the DEL molecule to the target protein, and covalent selections require only one selection round.

1.2.2 Solution-based DEL Selection Strategies

Solution-based DEL selection allows investigation of a target in its free state and eliminates the need for target immobilization. Various approaches utilizing different designs have been reported. The binder trap enrichment (BTE) strategy involves the integration of double-stranded DNA to the target

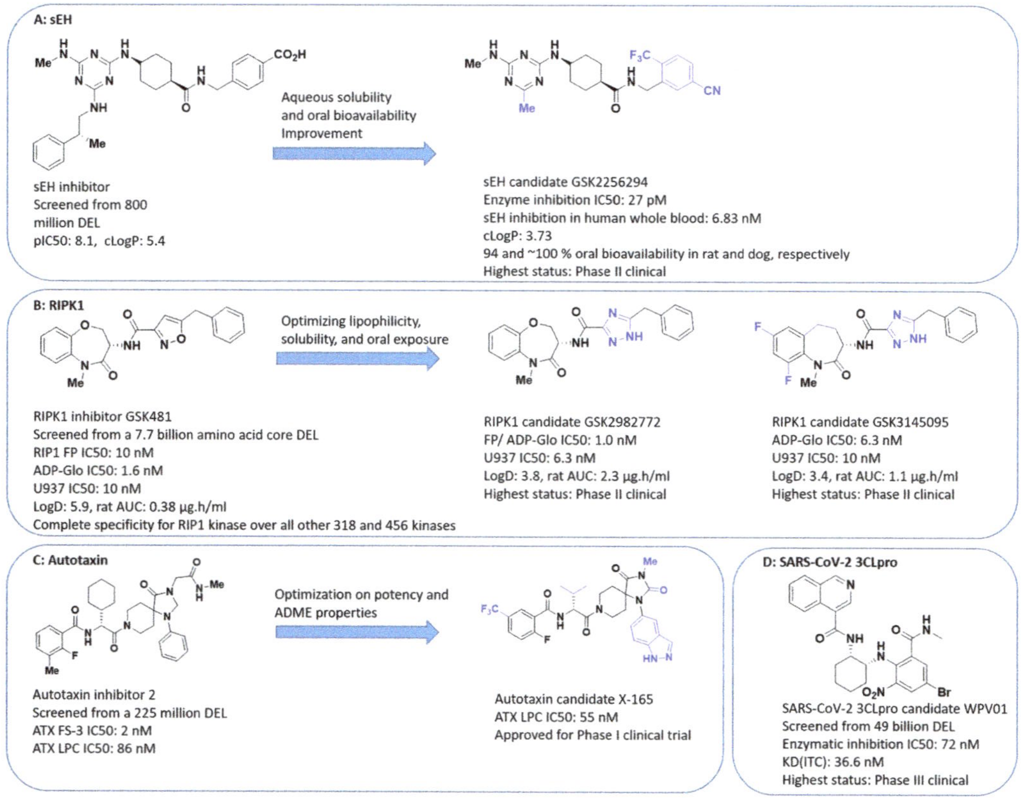

Figure 1.1 DEL-derived small molecules in various stages of clinical development. (A) The soluble epoxide hydrolase (sEH) candidate derived from DEL selections.[20,21] (B) Two receptor interacting protein 1 (RIP1) kinase candidates in clinical trials.[22–24] (C) DEL derived autotaxin inhibitor.[25] (D) Identification of the SARS-CoV-2 3CLpro candidate from DELs.[26]

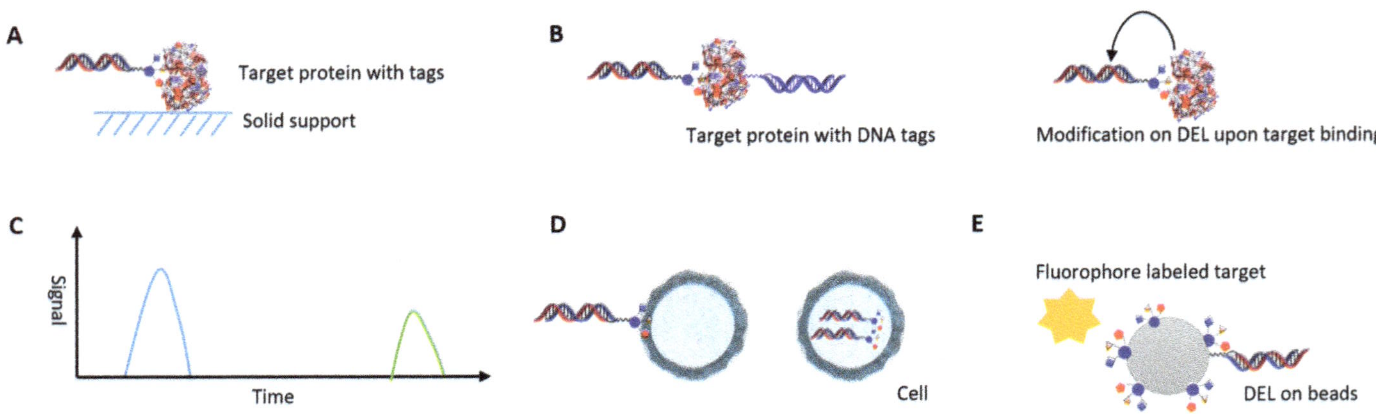

Figure 1.2 Different affinity-based selection strategies. (A) Protein-based selection strategy of using tagged protein and a solid support. (B) Solution-based selection strategy by using DNA tagged protein or modification of the DEL molecule upon protein binding. (C) CE- or LC-based DEL selection approach, where the retention time of the DEL–target complex is different from those of the unbound DEL and free protein. (D) Cell-based selection strategy for selection with cell surface targets or intracellular targets. (E) Fluorophore-based selection with DELs on beads.

protein, enabling the differentiation of bound DEL molecules through selective ligation of the DEL DNA to a DNA tag on the target protein within emulsion droplets.[31] The DNA tag on the target protein contains the primer binding site such that ligated constructs are selectively amplified in a subsequent PCR. Similarly, a single stranded DNA tag can be introduced to the target protein and the bound DEL molecule (as single stranded DNA) can be differentiated by complementary recognition and interaction dependent PCR (IDPCR).[32,33] Another variant of solution-based DEL selection is accomplished by utilizing a fusion protein on the target which can biotinylate or add poly A to the bound DEL molecules by induced proximity. Consequently, the DEL ligands can be enriched with corresponding streptavidin[34] or poly(dT)25 beads.[35] Solution-based selections can also be achieved by using a DNA-encoded dynamic library (DEDL) and PCR of bound DEL molecules upon target binding and photo-crosslinking (Figure 1.2B).[36] These methodologies alleviate the concern of a protein losing activity after immobilization, but specific instrumentation, DEL design, and customized selection methods may be required.

1.2.3 Capillary Electrophoresis (CE)- and Liquid Chromatography (LC)-based Selection Strategies

Separation methods based on capillary electrophoresis (CE) and liquid chromatography (LC) have been reported. A CE-based selection relies on the different electrophoretic rates of protein, free DEL, and the protein–DEL complex,[37,38] while an LC-based approach immobilizes protein on the resin and packs it into the column as a stationary phase (Figure 1.2C).[39–41] The chromatography-based strategies provide the advantages of screening compounds with different off-rates. However, due to intricate handling techniques and instrumentation requirements, these methods have not been widely implemented. Additionally, when compared to traditional formats, the screening capacity of CE and LC is significantly lower due to limitations in injection volumes and these methods need further optimization for adaptation to large scale DEL selections.

1.2.4 Cell-based DEL Selections

For targets that are difficult to prepare in a purified state, cell-based selections have been developed. This strategy maintains the target in its physiological condition, with the presence of cofactors and other interacting cellular components. However, the existence of additional proteins, cellular activities, and the expression level of the target can complicate the selection process, resulting in an elevated background and potentially diminished signals. Cell surface target-based selection employs centrifugation to distinguish between DEL binders and non-binders, eliminating the need for an affinity tag or solid support.[42–44] For selection against intracellular targets, DELs can be introduced into cells through conjugation with a CPP (cell penetrating peptide)[45] or *via* microinjection[46]/transfection to penetrate the cell membrane. The utilization of target proteins with a tag,[45] a fused prey protein,[46] or a target specific antibody[32] can facilitate the separation and extraction of DEL binders (Figure 1.2D). Further details regarding cell-based selection will be elaborated in a subsequent chapter (Chapter 3 – Live cell-based DEL selections).

1.2.5 DEL Selection with Bead-based DELs

In addition to the solution phase DELs used in the aforementioned methodologies, bead based DELs have been developed where each bead contains many copies of a single compound and thus commonly referred to as one bead, one compound (OBOC) DELs. OBOC DELs enable the separation process to be accomplished by probing the DEL beads with the target rather than immobilizing the target on a solid support. The selection of a DEL for binding to fluorophore-labelled RNA targets, has been accomplished with bead-based DELs with the employment of FACS (fluorescence-activated cell sorting) (Figure 1.2E).[47] In addition, bead-based DELs can be compartmentalized into droplets using microfluidics for functional screening.[48] The copy number of each compound on beads is much higher than the copy number in the solution phase DEL selection, and therefore this strategy augments the compound concentration and potentially increases the selection signal. Release of the compound from beads with compartmentalized droplets enables compound concentrations within typical screening concentrations (mid-micromolar), and traditional activity assays can be employed that require a binding equilibrium that is driven by the compound concentration (rather than the protein concentration). The selection process, however, is prolonged due to sorting and the throughput is significantly limited by sorting speed. This topic will be discussed in detail in another chapter (Chapter 2 – One-bead one-compound (OBOC) DELs for biochemical screens).

1.3 Protein Targets in DEL Selections

Some proteins are highly abundant in biological samples, allowing for the direct extraction and utilization of non-recombinant target protein in DEL selections. The absence of affinity tags can be circumvented by using cyanogen bromide (CNBr) or carboxylic acid beads, which immobilize target proteins by covalent attachment to amine groups on the protein. This immobilization method has been successful in identifying trypsin and albumin binders from DELs.[49] Alternative strategies, such as the chemical biotinylation of natural proteins, can also be employed, with subsequent immobilization achievable *via* streptavidin beads. Antibody or other interaction-based beads containing an affinity capture reagent to the target protein can also be utilized if available (Figure 1.3A).

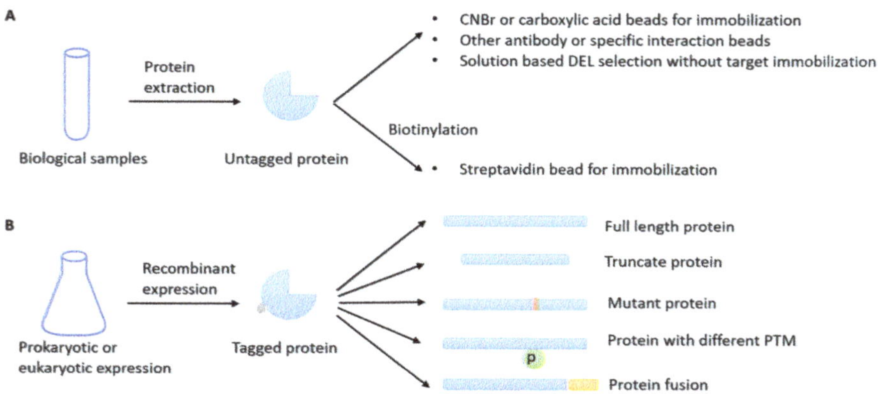

Figure 1.3 Different protein preparation strategies for DEL selection. (A) Target proteins directly extracted from biological samples. (B) Recombinant protein preparation for DEL selection.

Recombinant proteins are more frequently used for DEL selections than non-recombinant proteins, enabling the production of a construct with an appropriate affinity tag for use in selections and various downstream assays. The use of engineered, recombinant proteins allows for selections to be done against different domains of a target protein, the introduction of specific mutations to explore the binding pocket of interest or binding to a wild-type *versus* mutant form of a protein, and the confirmation of the mechanism of action (MOA) of hit compounds. Prokaryotic and eukaryotic expression systems can provide different post-translational modifications (PTMs) of targets, further aiding in the investigation of the impact of PTMs on protein status and DEL selection outcomes (Figure 1.3B). The ability to design and produce a wide range of proteins that span multiple target classes and represent different functional and/or structural states of potential drug targets has facilitated the successful application of DEL technology to many drug discovery programs.

1.3.1 Enzyme Ligands with Specific Mechanisms of Action (MOAs)

Enzymes are generally easy DEL targets owing to their catalytic nature, the ligandability of binding pockets, and in many cases a detailed understanding of structure and function. Various studies have been published utilizing DELs for enzyme ligand discovery.[3,50,51] While high throughput screening typically provides hits on the basis of enzymatic activity inhibition alone, DEL binding assays present unique advantages for identifying hits with a specific MOA, particularly for enzymes with multiple substrates. For enzymes with a compulsory ordered reaction mechanism, the binding of the first substrate alters the protein conformation, enabling the recognition of the second substrate. Many enzymes also catalyse reactions *via* a ping-pong mechanism, in which the second substrate can only bind after product release of the first substrate.[52] Therefore, inclusion of different substrates can reveal specific binding pockets and enable the investigation of the enzyme in different conformational and functional states. In a specific case study,[53] the target enzyme, Na-terminal acetyltransferase (Naa50), transfers an acetyl group from acetyl-CoA to acetylate the N-terminus of substrate proteins. Compared to the apo enzyme, inclusion of acetyl-CoA afforded additional signals from the selection. The hits were validated to have a specific mechanism that showed attenuated binding in the absence of acetyl-CoA. The DEL-derived ligands demonstrated potent inhibition of enzymatic activity and good engagement of intracellular Naa50. The binding mode was also explored in co-crystallization studies. The convenient inclusion of multiple DEL selection groups representing different functional and conformational states of the protein provides the opportunity for rapid investigation of a target protein at intermediate or unique conformations, thereby facilitating the identification of ligands with distinct MOAs.

In other case studies of kinases in DEL selections, Mer constructs varying in size and phosphorylation status, with and without ATP pre-incubation, along with three off-target proteins, were used in the DEL selection experiment. The variability in constructs and the presence of ATP facilitated the discernment of potential binding mechanisms. Selective inhibitors that displayed ATP-competitive and non-competitive features were identified in this study.[54] Additionally, the non-conserved cysteine located in or proximal to the active sites provides an effective strategy for kinase inhibitor development by offering enhanced potency or additional selectivity. In the case of Janus kinase 3 (JAK3), the non-conserved Cys909 yielded the selective covalent inhibitor ritlecitinib, making it an ideal target for covalent ligands.[55] Covalent-based DEL selection has been published on this target,[56] as well as for Bruton's tyrosine kinase (BTK)[57] and peptidyl-prolyl *cis*/*trans* isomerase NIMA-interacting-1 (Pin1).[56] The details of covalent DELs will be reviewed in another chapter (Chapter 6 – DELs with covalent warheads).

1.3.2 Protein–Protein Interaction (PPI) Targets

PPI targets are perceived as challenging, primarily due to the inherent nature of the interface of the protein pair. The PPI interface is both large and flat and presents unique challenges in targeting with small molecules. Peptoid libraries,[58] which effectively mimic the interaction of proteins, and the expansive chemical space provided by small molecule DELs both augment the probability of identifying small molecules directly. As noted in this study,[59] compounds with submicromolar potency in inhibiting lymphocyte function-associated antigen 1 (LFA-1) and intercellular adhesion molecule 1 (ICAM-1) were identified from selection of a 4.1 billion compound DEL against LFA-1.

Interleukin-2 (IL2), a cytokine that stimulates both regulatory T cells and effector lymphocytes, plays an important role in cancer immunotherapy and autoimmune diseases. IL2 interacts with its receptor IL2R, an interaction characterized by globular proteins, which has been proven challenging to inhibit.[60] Ligands of IL2 have been identified from different selections,[61,62] where the initial selection involving 30 000 compounds revealed a ligand with 2.5 μM K_d (dissociation constant), while the subsequent selection with a 669 240 member library yielded a ligand with 0.34 μM K_d, with the binding site located at the CD25 interface.

In another report,[58] an "anchor motif" strategy was utilized. Tryptophan was chosen as the starting point for design and synthesis of a peptoid DEL focused on tryptophan-mimetic indole side chains due to the presence of a key tryptophan residue in the "hot spot" of PPI. Using this focused library for the discovery of PPI inhibitors, TEAD4 binders were identified, exhibiting the ability to disrupt the hTEAD4–YAP interaction, and cellular activity was validated by reducing gene expression under control of the TEAD–YAP transcription factor complex.

1.3.3 G Protein-coupled Receptors (GPCRs)

GPCRs belong to a broad category of drug targets. GPCRs feature 7 transmembrane domains and respond to extracellular signals by triggering intracellular signalling pathways. Therefore, GPCRs are involved in a wide range of biological processes and are one of the most common classes of drug targets. Traditional GPCR screening has been primarily focused on cell based functional screening, structure based virtual screening and protein-based screening.[63] The initial application of DELs on GPCRs was hindered due to the challenges in producing high levels of recombinant GPCR protein in the appropriate conformation. However, cell-based selections were subsequently developed for directly screening GPCR ligands using cells instead of purified proteins. In this study,[42] the tachykinin receptor neurokinin-3 (NK3R) was highly overexpressed in HEK293 cells and screened with 15 billion compound DELs, yielding nanomolar antagonists. DEL selection conditions were further evaluated by conjugating the positive and negative ligands with DNA tags to investigate the correlation of selection signals and target expression density. The findings indicated that a high expression level of target protein is crucial for a successful DEL screen. Nonetheless, high density is difficult to achieve as the overexpression of recombinant GPCR protein may interfere with cell health and viability.

With progression of the understanding of GPCR structure and function, protein-based DEL selections against solubilized and purified GPCRs have been increasingly reported. In this study,[64] a protease-activated receptor 2 (PAR2) thermo-stabilized GPCR (StaR) was generated with enhanced stability in comparison to the wild-type counterpart. This protein retained the capacity to bind known PAR2 ligands. The cell-free DEL selection was performed against detergent solubilized PAR2, involving both apo protein and the pre-complexed form with the antagonist AZ6343, leading to the identification of nanomolar agonists and antagonists. In another study, purified human β2-adrenoceptor (β_2AR) was reconstituted in detergent-free, high-density lipoprotein (HDL) particles with the orthosteric site occupied by a high-affinity β-agonist, BI-167107. This strategy resulted in the identification of positive allosteric modulators (PAMs), which exhibited positive cooperativity with orthosteric agonists and high selectivity for β_2AR.[65] In addition to PAMs, negative allosteric modulators (NAMs) have also been identified for GPCRs by DEL selections. The detergent solubilized inactive μ-opioid receptor (μOR)/naloxone and active G-protein and agonist bound receptor were both included in the selection. A NAM compound of the μOR was identified, its structural mechanism was elucidated, and it was subsequently validated with *in vivo* efficacy.[66]

In addition to conventional bead-based affinity selection and cell-based selections, the receptor-affinity chromatography (RAC) approach was developed for ligand discovery with GPCRs. Using this strategy, the GPCRs were immobilized on a solid support, packed into stainless steel columns, and DEL selection was performed against the immobilized GPCRs incorporated in the HPLC (high-performance liquid chromatography) apparatus. Ligands for β_2AR, cysteinyl-leukotriene receptor (CysLT), angiotensin II type I receptor (AT1R) and endothelin receptor A (ETAR) have been identified *via* this selection strategy.[39–41]

Other strategies using GPCR-fusion protein constructs in cell-based DEL screens have been reported. The utilization of UltraID-δ opioid receptor fusion protein in Expi293F cells allowed the biotinylation of DEL molecules upon binding. Consequently, the selection could potentially improve recovery and enrichment of DEL ligands without necessitating the purification of the target protein.[34] A selective, G-protein-biased agonist of the K-opioid receptor was identified using live cell DEL selection by incorporating TurboID split protein and complementation.[67] A comprehensive review of cell-based DEL selection will be provided in another chapter (Chapter 3 – Live cell-based DEL selections).

1.3.4 DNA- and RNA-binding Proteins

DNA- and RNA-binding proteins primarily consist of transcription factors, DNA/RNA helicases, polymerases, nucleases, and scaffolding proteins. These proteins play an important role in regulating gene expression and maintaining genome integrity. DEL selection has been applied to various target types, but there is scepticism of using DELs for this class of proteins because of the DNA tags. The interference introduced by DEL DNA tags and DNA binding by the target potentially prevents the accurate recognition of small molecule binding, thereby hindering the effectiveness of the DEL selection. Despite the potential interference, successful applications of DEL selections to DNA- and RNA-binding proteins have been reported.

120 billion-compound DELs were screened against estrogen receptor α (ERα) WT and 3 gain-of-function (GOF) mutants. The resulting nanomolar ligand was conjugated with different E3 binder alkynes and further developed into degraders. The construct used in the DEL selection was the

ligand binding domain (LBD) of ERα with the DNA binding domain deleted. Therefore, the potential DNA binding issue in the DEL selection was circumvented.[68] Another DEL selection against the transcription factor TEAD-4 also utilized the same strategy by using a construct without the DNA binding domain.[58]

Monomeric RNase L has been utilized in DEL selection to identify dimer inducers and the compounds were further developed into RiboTACs (ribonuclease targeting chimeras) targeting the micorRNA-21 precursor.[69] The construct used in the DEL selection lacked DNA processing or binding, and therefore the DNA binding effect was not mentioned.

In addition to published cases, numerous conference posters and oral presentations have highlighted ligand discovery to DNA binding proteins using DEL selections. DEL selection has been implemented on these targets at HitGen as well. The overall DNA binding effect can be alleviated or differentiated from small molecule binding by using a variety of strategies. These include the use of different constructs, DNA binding site mutations, consensus DNA blocking, inclusion of counter targets, and ligand competition for specificity checks. Control DELs containing diversity in the DNA barcodes but lacking small molecule compounds can also be useful for discriminating DNA binding from small molecule binding. Furthermore, a bioinformatics analytical algorithm is crucial for false positive differentiation of binding based on DNA sequences.[70]

1.4 Nucleic Acid Targets in DEL Selection

In recent years, there has been a growing recognition of the pivotal role played by nucleic acids as druggable targets in the field of experimental therapeutics. The ability to selectively modulate the function and expression of nucleic acids has opened up new opportunities for the development of innovative therapeutic strategies. While DEL selection has been widely applied to a variety of protein targets, the utilization of DELs in nucleic acid targets remains questionable, owing to the concern that DEL DNA tags would interact with the nucleic acid target and interfere with screens designed to discover small molecule ligands. Various investigations have been applied to nucleic acid targets. The DEL selections utilized either targets labelled with biotin or poly A for immobilization and separation, or through targets labelled with a fluorophore for selection with bead-based DELs (Figure 1.4).

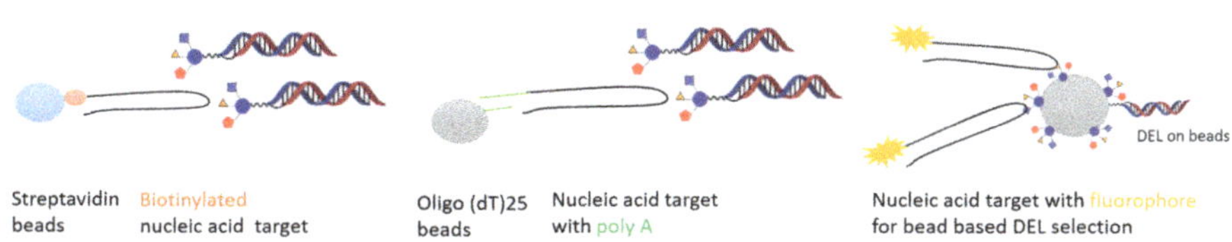

Figure 1.4 Different strategies for nucleic acid targets in DEL selection.

1.4.1 DNA Targets in DEL Selection

DEL selection against the G-quartet in the c-myc promoter has been reported using 33 distinct DELs containing 120 billion compounds. Positive binding compounds with selectivity over non-G-quartet DNA targets were identified. A parallel library with the small molecules cleaved off the DNA construct was included to help distinguish false positives due to DNA hybridization. Other strategies were not elaborated in this paper, which might be attributed to the established understanding that G-quartets possess a highly compact and stable structure.[71] This is the first report demonstrating DEL selection on a nucleic acid target, thereby encouraging more research for this application of DEL technology.

1.4.2 RNA Targets in DEL Selection

With the emergence of RNAs as therapeutic targets, DEL selection on RNA targets has been explored more intensively. A published study reported the use of a bead-based DEL to screen a library of RNA structures. RNA structures were labelled with fluorophores and sorted by flow cytometry. A base paired counter screen with a different fluorophore label was incorporated to increase selectivity and minimize false positives (two-color FACS). The paper did not provide specific details on any DNA binding interference, which may be attributed to the use of a solid-phase library in the selection process. In contrast to the solution phase DEL, the stoichiometry of DEL DNA tags and small molecules in bead-based libraries is typically not 1 : 1, with the small molecule in excess of the DNA. The utilization of an RNA library with a randomized 3×3 region instead of a specific RNA target helped alleviate DNA binding interference. The screen identified a ligand that was active in cellular assays for a specific RNA structure present in primary microRNA-27a.[47] This strategy using a specific RNA target has been reported in another study where the RNA repeat expansion r(CUG)exp was used in the two-color FACS screening. Hits identified were further augmented when attached to an RNA degrading molecule.[72]

Solution phase DEL presents more obstacles in DEL selection against RNA targets. Given the 1 : 1 stoichiometry of DNA tags and small molecules in most DELs, the potential effect of DNA tags on RNA structure and challenges associated with DNA binding have been difficult to tackle. This is especially concerning for less compact and stable RNA targets, which usually adopt dynamic conformations under different conditions. A comprehensive exploration of solution phase DEL selection on RNA targets has been reported.[70] The utilization of RNA patches, which are short fragments of the RNA target and serve as surrogates to hybridize with DEL DNA tags, in combination with competitive elution, significantly reduce DEL DNA tag binding and increase the specificity of the DEL selection. In addition, informatics filtering tools such as K-mer analysis and motif search were developed to differentiate DNA binding computationally. This approach has led to the identification of specific ligands for FMN riboswitch RNA from the selection.[70] A separate instance of DEL selection against an RNA aptamer target has been noted, but the specific details of the selection process were not disclosed.[73]

1.5 Outlook

The intrinsic nature of DEL design and screening, whereby libraries of small molecules are tethered to a second moiety *via* a linker and screened for target binding, makes DELs an ideal technology for PROTAC (proteolysis-targeting chimera) applications. This can be achieved by separate screening on E3 ligases or POIs (proteins of interest) for ligand identification and further development of PROTAC *via* conjugation. Moreover, the direct discovery of PROTACs in ternary complex formation based on the selection format has been reported.[74–76] This application is discussed in depth in another chapter (Chapter 7 – Application of DELs for E3 ligase ligand and PROTAC development). Given the large chemical space, high sensitivity and fast investigation of structure–activity relationships (SAR), DELs can also be employed in the discovery of molecular glues. A compound with a molecular glue mechanism of action may require a precise and nuanced interaction with a target protein pair. The DNA tag and linker will potentially contribute to binding of DEL compounds to the target protein pair, a possibility that needs to be meticulously evaluated in the characterization of compounds derived from molecular glue-based DEL screens. In addition, the conjugation nature of DELs and the expansive chemical space render them a beneficial tool for antibody–drug conjugate (ADC) development. DEL selection can be utilized for the identification of payloads for specific targets. The conjugation of DEL-derived molecules to an antibody has been reported; in this particular case, the tumour antigen-specific DEL binders were developed as a switch for CAR T (chimeric antigen receptor T) cell cytotoxicity and specificity.[77]

Despite the success in ligand discovery for numerous target classes, there is still failure in getting hits for certain target types. Intrinsically disordered proteins (IDPs) lack a fixed or an ordered three-dimensional structure under physiological conditions, and they adopt different conformations with dynamic changes. The high flexibility and the lack of well-defined binding pockets make ligand discovery on these proteins challenging. To our knowledge no successful IDP DEL selection case study has been published. Developing DELs covering a larger and more diverse chemical space, or new selection techniques or strategies may be necessary for successful application of DELs to IDPs. Transporters and ion channels are multi-transmembrane proteins involved in various biological processes and constitute a significant portion of drug targets. The lack of DEL selection reports on these target types may be due to the difficulty in preparing the recombinant proteins as well as challenges in achieving good selectivity over closely related family members. Considering these challenges, more efforts are required to further investigate the utilization of DEL selection on these target classes. This would entail not only overcoming the technical difficulties associated with the protein preparation but also developing innovative strategies to enhance the selectivity and efficacy of DEL selections.

In addition to advances in DEL chemistries and refinement of selection strategies, we anticipate that the further development of NGS technologies will significantly impact the field. The progression of NGS is expected to reduce the cost and substantially enhance sequencing capacity. This improvement may bring about a paradigm shift in DEL selection practices, by potentially changing the selection input and the number of selection rounds significantly. For example, strategies to determine compound affinity from selection data[14,15] will be facilitated by progression of NGS technology. With current practice, binding affinity can be inferred from DNA sequence counts of enriched compounds, but for a number of reasons a strict correlation of DNA sequence counts with compound affinity does not exist. The ability to precisely determine compound potency from primary DEL selection data would represent a significant improvement of the technology. Investigation on data interpretation has been reported,[16–19] but with the larger sequencing datasets generated with the high NGS capacity in the future, the analytical algorithms employed to process and interpret the DEL selection results will also require enhancements in both efficiency and accuracy.

Machine learning (ML) approaches have been increasingly employed in interpreting and denoising DEL selection data.[78] Furthermore, they have significantly contributed to the discovery of ligands for a variety of targets.[79,80] Continued development and refinement of ML models, combined with the large quantity of data provided by DEL selection, are expected to considerably reduce the associated cost and optimize the efficiency of the DEL process. In turn, this should enhance the success rate and overall impact of the DEL platform for hit discovery.

References

1. S. Brenner and R. A. Lerner, *Proc. Natl. Acad. Sci. U. S. A.*, 1992, **89**, 5381.
2. Z. J. Gartner, B. N. Tse and R. Grubina, *Science*, 2004, **305**, 1601.
3. M. A. Clark, R. A. Acharya and C. C. Arico-Muendel, *Nat. Chem. Biol.*, 2009, **5**, 647.
4. L. Mannocci, Y. Zhang and J. Scheuermann, *Proc. Natl. Acad. Sci. U. S. A.*, 2008, **105**, 17670.
5. F. Buller, M. Steiner and J. Scheuermann, *Bioorg. Med. Chem. Lett.*, 2010, **20**, 4188.
6. Z. Zhu, A. Shaginian and L. C. Grady, *ACS Chem. Biol.*, 2018, **13**, 53.
7. D. T. Flood, X. Zhang and X. Fu, *Angew. Chem., Int. Ed.*, 2020, **59**, 7377.
8. G. Yang, D. He and Y. Zhu, *Bioconjugate Chem.*, 2021, **32**, 667.
9. P. Cai, G. Yang and L. Zhao, *Org. Lett.*, 2019, **21**, 6633.
10. Q. Chen, X. Cheng and L. Zhang, *SLAS Discovery*, 2020, **25**, 523.

11. A. Sannino, E. Gabriele and M. Bigatti, *ChemBioChem*, 2019, **20**, 955.
12. S. Puglioli, S. Oehler and L. Prati, *Chem. Sci.*, 2023, **14**, 12026.
13. T. L. Foley, W. Burchett and Q. Chen, *SLAS Discovery*, 2021, **26**, 263.
14. Q. Chen, J. Hall and T. L. Foley, *Biochem. Biophys. Res. Commun.*, 2020, **533**, 249.
15. J. Hall, T. L. Foley and Q. Chen, *Biochem. Biophys. Res. Commun.*, 2020, **527**, 250.
16. L. Kuai, T. O'Keeffe and C. Arico-Muendel, *SLAS Discovery*, 2018, **23**, 405.
17. A. L. Satz, R. Hochstrasser and A. C. Petersen, *ACS Comb. Sci.*, 2017, **19**, 234.
18. P. Kómár and M. Kalinić, *ACS Comb. Sci.*, 2020, **22**, 410.
19. B. Chen, M. M. Sultan and T. Karaletsos, *J. Chem. Inf. Model.*, 2024, **64**, 1123.
20. S. L. Belyanskaya, Y. Ding and J. F. Callahan, *ChemBioChem*, 2017, **18**, 837.
21. P. L. Podolin, B. J. Bolognese and J. F. Foley, *Prostaglandins Other Lipid Mediators*, 2013, **104–105**, 25.
22. P. A. Harris, B. W. King and D. Bandyopadhyay, *J. Med. Chem.*, 2016, **59**, 2163.
23. P. A. Harris, S. B. Berger and J. U. Jeong, *J. Med. Chem.*, 2017, **60**, 1247.
24. P. A. Harris, J. M. Marinis and J. D. Lich, *ACS Med. Chem. Lett.*, 2019, **10**, 857.
25. J. W. Cuozzo, M. A. Clark and A. D. Keefe, *J. Med. Chem.*, 2020, **63**, 7840.
26. N. Hou, L. Shuai and L. Zhang, *ACS Cent. Sci.*, 2023, **9**, 217.
27. Q. Chen, in *DNA-Encoded Chemical Libraries*, ed. D. Israel and Y. Ding, Springer, New York, NY, 2022, ch. 19, pp. 155–164.
28. W. Decurtins, M. Wichert and R. M. Franzini, *Nat. Protoc.*, 2016, **11**, 764.
29. Z. Zhu, L. C. Grady and Y. Ding, *SLAS Discovery*, 2019, **24**, 169.
30. H. Ma, J. B. Murray and H. Luo, *RSC Med. Chem.*, 2022, **13**, 1341.
31. L. K. Petersen, P. Blakskjær and A. Chaikuad, *MedChemComm*, 2016, **7**, 1332.
32. L. M. McGregor, T. Jain and D. R. Liu, *J. Am. Chem. Soc.*, 2014, **136**, 3264.
33. L. M. McGregor, D. J. Gorin and C. E. Dumelin, *J. Am. Chem. Soc.*, 2010, **132**, 15522.
34. B. Cai, A. B. Mhetre and C. J. Krusemark, *Chem. Sci.*, 2023, **14**, 245.
35. L. A. Schneider, B. Sauter and K. Dagher, *J. Am. Chem. Soc.*, 2023, **145**, 20874.
36. Y. Zhou, W. Shen and J. Peng, *Bioorg. Med. Chem.*, 2021, **45**, 116328.
37. J. Bao, S. M. Krylova and L. T. Cherney, *Anal. Chem.*, 2016, **88**, 5498.
38. S. Kochmann, A. T. H. Le and R. Hili, *Electrophoresis*, 2018, **39**, 2991.
39. Q. Liang, H. Zuo and T. Yang, *Eur. J. Med. Chem.*, 2022, **233**, 114212.
40. Q. Liang, X. Zhao and X. Fu, *Bioorg. Chem.*, 2021, **112**, 104986.
41. Q. Liang, J. He and X. Zhao, *J. Med. Chem.*, 2021, **64**, 4196.
42. Z. Wu, T. L. Graybill and X. Zeng, *ACS Comb. Sci.*, 2015, **17**, 722.
43. S. Oehler, M. Catalano and I. Scapozza, *Chemistry*, 2021, **27**, 8985.
44. Y. Huang, L. Meng and Q. Nie, *Nat. Chem.*, 2021, **13**, 77.
45. B. Cai, D. Kim and S. Akhand, *J. Am. Chem. Soc.*, 2019, **141**, 17057.

46. L. K. Petersen, A. B. Christensen and J. Andersen, *J. Am. Chem. Soc.*, 2021, **143**, 2751.
47. R. I. Benhamou, B. M. Suresh and Y. Tong, *Proc. Natl. Acad. Sci. U. S. A.*, 2022, **19**, e2114971119.
48. A. B. MacConnell, A. K. Price and B. M. Paegel, *ACS Comb. Sci.*, 2017, **19**, 181.
49. L. Mannocci, S. Melkko and F. Buller, *Bioconjugate Chem.*, 2010, **21**, 1836.
50. L. H. Yuen, S. Dana and Y. Liu, *J. Am. Chem. Soc.*, 2019, **141**, 5169.
51. Y. Ding, H. O'Keefe and J. L. DeLorey, *ACS Med. Chem. Lett.*, 2015, **6**, 888.
52. R. A. Copeland, in *Evaluation of Enzyme Inhibitors in Drug Discovery*, ed. R. A. Copeland, Wiley, Hoboken, New Jersey, 2005, ch. 2, pp. 21–47.
53. P.-P. Kung, P. Bingham and B. J. Burke, *ACS Med. Chem. Lett.*, 2020, **11**, 1175.
54. J. W. M. Nissink, S. Bazzaz and C. Blackett, *J. Med. Chem.*, 2021, **64**, 3165.
55. H. Xu, M. I. Jesson and U. I. Seneviratne, *ACS Chem. Biol.*, 2019, **14**, 1235.
56. L. Li, M. Su and W. Lu, *ACS Med. Chem. Lett.*, 2022, **13**, 1574.
57. J. P. Guilinger, A. Archna and M. Augustin, *Bioorg. Med. Chem.*, 2021, **42**, 116223.
58. V. B. K. Kunig, M. Potowski and M. Akbarzadeh, *Angew. Chem., Int. Ed.*, 2020, **59**, 20338.
59. C. S. Kollmann, X. Bai and C.-H. Tsai, *Bioorg. Med. Chem.*, 2014, **22**, 2353.
60. D. E. Scott, A. R. Bayly and C. Abell, *Nat. Rev. Drug Discovery*, 2016, **15**, 533.
61. M. Leimbacher, Y. Zhang and L. Mannocci, *Chemistry*, 2012, **18**, 7729.
62. A. Gironda-Martínez, É. M. D. Gorre and L. Prati, *J. Med. Chem.*, 2021, **64**, 17496.
63. P. Kumari, E. Ghosh and A. K. Shukla, *Trends Mol. Med.*, 2015, **21**, 687.
64. D. G. Brown, G. A. Brown and P. Centrella, *SLAS Discovery*, 2018, **23**, 429.
65. S. Ahn, B. Pani and A. W. Kahsai, *Mol. Pharmacol.*, 2018, **94**, 850.
66. E. S. O'Brien, V. A. Rangari and A. El Daibani, *Nature*, 2024, **631**, 686.
67. B. Cai, A. E. Daibani and Y. Bai, *JACS Au*, 2023, **3**, 1076.
68. J. S. Disch, J. M. Duffy and E. C. Y. Lee, *J. Med. Chem.*, 2021, **64**, 5049.
69. S. M. Meyer, T. Tanaka and P. R. A. Zanon, *J. Am. Chem. Soc.*, 2022, **144**, 21096.
70. Q. Chen, Y. Li and C. Lin, *Nucleic Acids Res.*, 2022, **50**, e67.
71. A. Litovchick, X. Tian and M. I. Monteiro, *Molecules*, 2019, **24**, 2026.
72. Q. M. R. Gibaut, Y. Akahori and J. A. Bush, *J. Am. Chem. Soc.*, 2022, **144**, 21972.
73. H. Mukherjee, J. C. Blain and L. E. Vandivier, *ACS Chem. Biol.*, 2020, **15**, 2374.
74. Q. Chen, C. Liu and W. Wang, *ACS Chem. Biol.*, 2023, **18**, 25.
75. J. W. Mason, Y. T. Chow and L. Hudson, *Nat. Chem. Biol.*, 2024, **20**, 170.
76. S. Liu, B. Tong and J. W. Mason, *J. Am. Chem. Soc.*, 2023, **145**, 23281.
77. A. V. Stepanov, J. Xie and Q. Zhu, *Nat. Biomed. Eng.*, 2024, **8**, 529.
78. C. Zhang, M. Pitman and A. Dixit, *J. Chem. Inf. Model.*, 2023, **63**, 5120.
79. K. McCloskey, E. A. Sigel and S. Kearnes, *J. Med. Chem.*, 2020, **63**, 8857.
80. S. Ahmad, J. Xu and J. A. Feng, *J. Med. Chem.*, 2023, **66**, 16051.

CHAPTER 2

DNA-encoded Library Machine Learning Applications

ERIC A. SIGEL

Signel IX LLC, Belmont, MA 02478, USA
Email: signelix@gmail.com

2.1 Machine Learning in Drug Discovery

Machine learning (ML) has held promise to the drug discovery community for the last several decades. Until recently, applications of ML approaches in small molecule drug discovery fell under the discipline of quantitative structure–activity relationship (QSAR) analysis, often practiced in computational chemistry teams.[1] With AI/ML popularity at its peak, ML and data science have emerged as specialized fields distinct from traditional computational chemistry, adapting to the increasing scale and complexity of algorithms, data logistics, and prediction spaces. The increased focus and rapid growth in this space, beyond the hype-cycle that has traditionally heavily impacted interest in AI/ML both positively and negatively, is driven by several major factors: scope and scale of data, access to computational resources, and improvement of and insight into novel machine learning approaches. While, in the general application of ML, the data explosion has occurred as a result of social media and the internet of things (IoT), the data growth in drug discovery and diagnostics has been driven largely by the increasing capability and subsequent commoditization of DNA sequencing. In the past two decades, the volume of sequencing has risen massively with a concomitant inverse shift in cost, for example, leading to whole genome sequencing availability in the low \$100s. This increase in large-scale readouts has made generating billions of data points easily affordable and

Drug Discovery Series No. 85
DNA-encoded Library Technology for Drug Discovery
Edited by Guansai Liu, Casey J. Krusemark and Jin Li

Published by the Royal Society of Chemistry, www.rsc.org

commoditized across the industry and academia. Where genomic data and analysis has led the pack, multi-omics has rapidly followed behind driving growth in transcriptomics, proteomics, genomics, epigenomics, and microbiomics as well as driving the more recent shift to single-cell multiomics.[2]

For chemistry and chemical biology however, this has been a longer process. Data science enabling volumes of data in the chemical biology space has been much slower to ramp up due to a number of significant cost, resource and time barriers. (1) Synthesis of bespoke molecules is commonly estimated to cost in the low to mid-single digit thousands of dollars,[3] requiring the financial resources (and years of historical synthesis) of a large pharmaceutical company to create small molecule collections in the low millions of molecules. (2) Testing these molecules requires substantial investment in robotics, protein synthesis, and assay miniaturization and considerable protein synthesis budgets for high-throughput screening (HTS) runs. All told, investment requires millions of dollars to outsource, and billions of dollars to generate robust in-house collections, which puts this capability out of reach for many biotechs and academics.

A number of directions have emerged to create the data needed to generate ML-enabling inputs for small molecule applications. Companies such as Wuxi AppTec and Evotec provide modest collections in hundreds of thousands of molecules for screening. Affinity selection mass spectrometry (ASMS) has emerged as a cost-effective approach for data generation,[4,5] and aggregated data sets including public (*e.g.* ChEMBL) and commercial (*e.g.* GoSTAR) have been used extensively in the industry and for academic machine learning efforts, *e.g.* ref. 6 and 7.

DNA-encoded libraries (DELs) have provided an alternative, economically feasible approach by effectively co-opting the revolution in next generation/high throughput sequencing and bringing its well-established data generation engine to bear on chemical biology data, albeit as a binding assay rather than an activity assay for the most common DEL applications. In fact, DNA-encoded library technology development benefitted from serendipitous timing in that 454 Life Sciences had begun to gain traction in the early 2000s when the early industrial DEL projects had been initiated. This shift to high-throughput sequencing rapidly ramped data acquisition from Sanger sequencing (maxed out at hundreds of reads) to read counts into tens of thousands of reads. With Illumina, Inc. entering into the picture later in the decade, data generation rapidly moved to tens and ultimately hundreds of millions of reads per lane by the first half of the 2010s.

Despite having broken the data generation boundary in the late 2000s to early 2010s, the application of machine-learning to DEL has taken a number of years to start gaining traction. Some key factors that have slowed the progress of this approach include the needed development and adoption of novel ML techniques, the cost of model validation, the human elements, *e.g.* 'Not Invented Here', and the complexity of DELs and data output.

As reflected above, AI/ML has gone through a number of hype cycles and has suffered from a 'bad reputation' until recent notable successes have accelerated its adoption. With improving capabilities and approaches, such as deep learning and graph convolutional neural networks (GCNNs),[8,9] the application of AI/ML to molecular data has begun to show substantial progress. This advancement enables the generalization of model predictions beyond the extensive, yet more series-focused, QSAR efforts[10] that have dominated ML applications in the past decades. Instrumental work[6,11] in the mid/late-2010s demonstrated that molecular predictions can be used across a number of drug development tasks, *e.g.* ref. 12 and 13.

A further road-block that continues to limit the broad application of predictive models in small molecule space in general is the high cost to validate these models in terms of the synthetic cost of generating molecules, the resource cost of testing those molecules, and the time investment needed to get to a validation point. In other words, where DEL data have driven down the cost of input data and ML approaches have reached a point where their application is fairly straightforward, DEL-ML and other small molecule predictive approaches[14] have reached the next impasse to be solved, *i.e.* how do we know that the model is working; do we spend the resources to test the model; and is the time frame needed to test these models going to produce a result of value. These complex factors interplay with the human factor, *i.e.* traditional heavy skepticism of ML by chemists and biologists and the understandable tendency to focus on the tried-and-true approaches as a better use of limited resources. At issue, however, is that a novel technology (ML-DEL and ML for drug discovery in general), requiring considerable learning and development to apply it effectively in the broader discovery context, is being held up to the standard of achieving as good or better results than traditional approaches that have been optimized for several decades. In fact, instead of being encouraged that ML approaches hold their own against other hit-finding approaches that cost more and take significant time, a common reaction appears to be companies taking an implicit or explicit 'one-and-done' mindset. The overall effect is to compound slow timelines for validating models even further with the goal to get the experiment "as perfect as possible". Accordingly, teams end up in long-running discussions to answer questions: how to run the initial DEL experiment? What constitutes a strong enough retrospective validation to move forward? Deliberation over the ideal list of molecules to order adds months to the inherent 3-to-12-week molecule acquisition time frames, and inevitable delays in molecule testing. All told, these planning, logistics and operational time costs create a dynamic that feeds a conclusion that ML in drug discovery approaches is far too slow to impact the early drug discovery process in contrast to traditional med-chem approaches advancing one-off molecules through a number of cycles before the ML has even read out.

DEL as a technology in addition to providing an extremely large source of data for learning also creates a number of challenges. Well-known[15] DEL data considerations include the difficulty in normalization of the number of molecules per library instance, the presence of promiscuous and matrix binders, multiple protein conformations in selections, the presence of multiple binding sites and binding modes, the contribution of the attached DNA to the binding signal, and reaction yields and alternative products. With traditional analysis, this can manifest as the failure of enriched molecules to produce signal in off-DNA synthesis. The impact with respect to ML of the 'messy' nature of DEL data creates both opportunities and challenges. ML being in many cases a statistical analytical approach that can identify 'hidden' data trends, obscured by this complexity from the analyzing scientist, provides a mechanism to holistically 'understand' the data. This broader understanding of the data in aggregate has the potential to provide the tools for teasing out the signal from the noise, and, once further advanced as a technology, the potential to evaluate multi-modal binding occurring within a single experiment. On the other hand, there is a balance to be had with the 'garbage-in-garbage-out' potential created by highly confounding data sets. Are there spurious trends that are created due to uninteresting binding events, *e.g.* promiscuous binding? Are trends being lost due to library chemical make-up, inconsistency of reactions during library synthesis, and the inherent variability in read-outs?

To improve DEL analysis with respect to these several challenges, there have been significant developments. DEL-ML, and predictive drug discovery in general, permits a shift in mindset in discovery efforts (Figure 2.1 provides a schematic of a basic DEL-ML workflow). Prediction allows the model user to focus the model on molecular space that an HTS, structure based drug design (SBDD) or a traditional DEL approach does not address. The predictive approach allows the "dialing-in" of the key properties of the molecules that are in the targeted prediction collection. Such properties can include drug-likeness, ADME/Tox predictions, *etc.* However, the target properties can also include practical considerations such as the cost per molecule, synthetic accessibility, and the speed of acquisition or synthesis. While an HTS approach rapidly narrows the chemistry of interest to a small number of molecules, prediction has the potential to open up a much broader list of scaffolds. Led by Enamine's REAL collection with others (Wuxi GalaXi, Mcule Ultimate), the advancement of predictive approaches is a timely development to capitalize on the unprecedented scale of molecules that can now be synthesized at a reasonable cost and time-scale. In our publication in 2020 on DEL-ML,[3] we demonstrated the cost shift enabled by ML with the expense to acquire predicted molecules dropping by an estimated 25- to 250-fold per molecule as compared to the estimated cost per molecule using our DEL off-DNA follow-up approach. With the ability to test hundreds to thousands of molecules for the cost of tens of bespoke syntheses, a number of benefits arise: (1) more "shots on goal" lead to a greater diversity of hits, (2) better understanding of model performance can be gained with larger, more statistically relevant data sets and (3) the data produced are in and of themselves at a scale that can start to inform the next model. With the vast increase in accessible, low-cost chemical spaces now

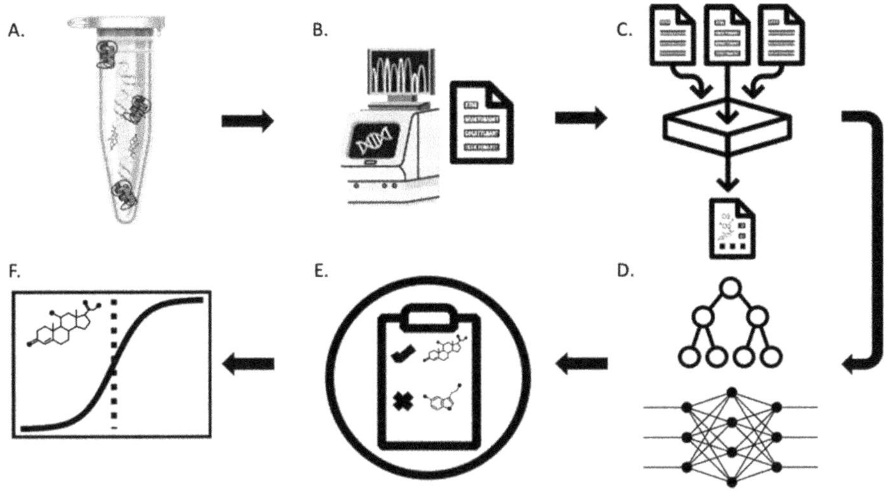

Figure 2.1 A schematic representation of the DEL-ML workflow for hit-identification. (A) A standard DEL selection is performed incubating the protein with the full DEL. Multiple conditions including a no target control will be performed in parallel. Washes are performed to remove non-binders. (B) Next-generation sequencing (NGS) of the retained DEL members is used to generate hundreds of millions of data points. (C) The NGS data are processed in order to translate sequences to the encoded chemistry and to counts for each encoding. (D) Data are further analyzed to classify molecules as positive or negative examples. The classified data are used to generate machine learning models, *e.g.* random forest (RF) or graph convolutional neural networks (GCNNs). (E) The models are applied to molecule collections to generate a prediction score for each molecule. These collections may be in-house HTS collections, virtual libraries, or commercial collections such as Enamine REAL that can reach billions of molecules in size. (F) A diverse set of molecules with high prediction scores are selected for purchase or synthesis and tested in a binding or activity assay.

reaching tens to low hundreds of billions of molecules, and the application of DEL-ML models in conjunction with the state-of-the-art multi-parameter optimization (MPO)-oriented predictions has the potential to provide increasingly de-risked and diverse starting points that can enable progression of better discovery programs.

The data complexity problem has received a fair amount of attention in recent years with a number of articles focusing on DEL-denoising, improved library design, and QC.[15–19] There is a balance to be struck between data clean-up and the strong 'push' in the data science world to 'let the ML learn how to handle the data complexity'. In our proof-of-concept paper,[3] we adopted a disynthon aggregation approach for our models, motivated by the idea that increased sequencing depth and statistical analysis over two library cycles would enhance the signal-to-noise ratio (as discussed further below). Our use of this disynthon approach was not our initial choice for the

Google–X-Chem collaboration, and in fact our earliest passes fed the full instance molecules as the input. However, in these early attempts, the model metrics on testing and validation did not prove strong enough to motivate molecule acquisition and assay. Moving to the disynthon approach, we rapidly achieved stronger metric read-outs for the models and ultimately justifying the move forward to test predicted molecules. The take-home message in this result should not be that a full-molecule is inferior, but rather that the many decisions required to drive instance-level modeling are more complex in comparison to the more robust disynthon approach. Notably, the Google and PhiloChem teams recently published on the use of instance-level models with success.[20]

2.2 Data Processing and Modeling Considerations in DEL-ML

DEL data provide a rich source of information for machine learning. As with most machine learning applications, how the data are processed can significantly affect the results. The key elements of data analysis include (1) how to analyze data for use as input, *e.g.* classification *vs.* regression-oriented data modeling and decisions of how to classify inputs as positive and negative; (2) molecular representations can be varied including traditional fingerprints (Morgan/ECFP, FCFP, ECFC, chemical fingerprints, MACCs keys, *etc.*), learned representations (GCNN embeddings or auto-encoder derived fingerprints); (3) algorithms ranging from tree-based to neural networks and deep learning; (4) the use of multiple experimental conditions; (5) selection approaches for inclusion of negative data from the DEL experiment and from other reference sources (public data, sampled targeted prediction collections); (6) test and validation set splitting approaches and (7) metrics for assessing model performance. With this wide range of factors, the model possibilities for DEL-ML predictions rapidly combine to create a large matrix of modeling approaches. The current state of literature in this space does not yet produce a definitive "winning" approach for DEL-ML and, as such, analysis across many different approaches is warranted. Figure 2.2 shows a proposed pipeline to enable the combinatorial generation of model possibilities.

2.3 Preparation of Data for Input

For classification models using DEL data, the modeler must provide a definition for what data constitute positive training examples, or the ML approach must learn where to apply such a cutoff. For regression models, this cutoff is not required, as the model can be trained to learn, for example, the number of counts expected for a given instance. With such regression models, the learning is subject heavily to the variability in sequencing and library constitution. For classification models, a metric of enrichment can be defined and then applied across multiple conditions. For example, a positive cutoff may include high enrichment in the target condition, low or no

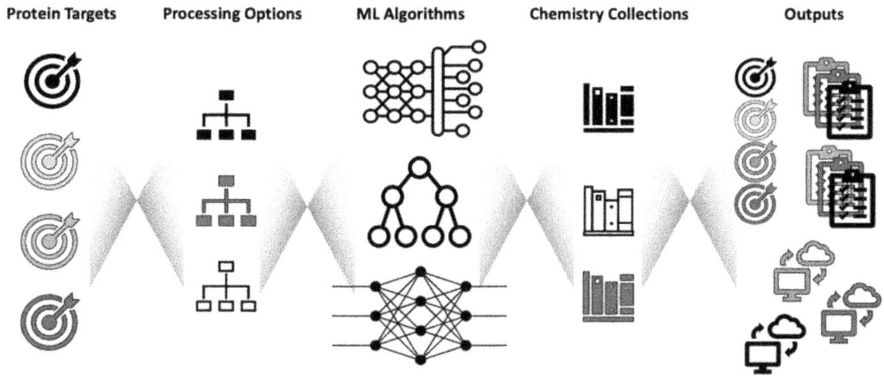

Figure 2.2 A computational pipeline for generating DEL-ML models. The proposed setup produces a combinatorial expansion for training on DEL data sets. DEL selection data from multiple targets, multiple conditions, or multiple target constructs serve as the input. The data are then processed using a range of options, for example, multiple chemical fingerprint representations may be calculated and multiple positive/negative classification definitions may be used specifying different cut-offs and/or combinations of selection conditions. The classified data are then used in conjunction with a variety of algorithms (RF, XGBoost, GCNNs, *etc.*) to generate multiple models. Models are assessed for performance using several prediction metrics against test or validation data. The most promising are applied to a number of chemical collections which may include virtual libraries, commercial collections, or internal HTS collections. The output from each model-collection prediction run is a list of molecules selected to meet filter criteria (*e.g.* Lipinski's rule of five) and diversity criteria in order to identify a broad range of potential chemical series. The use of cloud resources provides the opportunity for substantial parallel computation at multiple steps.

enrichment in a no-target control and low or no enrichment in a target plus competitor condition where the competitor is known to bind in the binding pocket/binding mode of interest. Such cutoff conditions can be arbitrarily complex but require the consideration of over-specifying data leading to a reduction in the training set, and the potential for eliminating signal that may not meet specific criteria but is still useful for learning relevant binding chemical motifs.

In addition to using DEL signal in the specific experiment, information on DEL binding of molecules across many targets can be used to develop a promiscuity metric to reduce the learning of chemical motifs that are non-specific for a target. A practical consideration in using this approach, however, includes the degree of access that practitioners of DEL-ML have to such historical data in order to enable such a filter. Further, some care must be taken in applying promiscuity filters not to throw out valuable signal. Kinase domains, for example, are known to prefer common elements (*e.g.* hinge binding structures) that are relevant broadly across the kinase subclass.

Removal of common chemical motifs should take these subtleties into consideration or risk removing important pattern information. An alternative, promising direction in defining promiscuous and matrix binding capitalizes on the large volume of DEL data across a range of targets to develop a ML-based filter. Filtering using such ML-based frequent binder models has the potential for improved removal of less desirable molecules and false positives when used in both traditional and ML-based DEL data analyses.

2.4 Molecular Representation

Structural representation has gone through a number of iterations and refinements in the past decades moving from keys tabulating the presence or absence of structure to topological fingerprints (that capture linear and more recently radius defined substructures in molecules) to three-dimensional fingerprints, and increasingly to learned fingerprints.[6,21] Such learned fingerprints can be produced either directly through the ML mechanism for the specific task or by the use of separate tasks (*e.g.* variational autoencoders (VAEs)) to generate fingerprints that are specific either to the data set and task or to a defined set of molecules that are of interest to the prediction process (*e.g.* VAE fingerprint generated using the target collection). In the case of ECFP/Morgan fingerprints, chemical fingerprints, MACCs keys and other calculated fingerprints, off-the-shelf tools, both commercial (ChemAxon, *etc.*) and open source (RDKit),[22] can be used to generate the fingerprint representation. For these tools, parameterized decisions are made upfront including the size of the fingerprint (1024, 2048, *etc.*) and the use or non-use of feature counts. In our work,[23] preliminary results suggest that low order fingerprint sizes (*e.g.* MACCs keys) lead to inferior results across many model types when trained on the same data sets. However, the question remains open as to what fingerprint information density is ideal. Further, as this is currently assessed using model performance metrics rather than prospective molecular assay results, considerable work remains to understand how best to optimize even simple fingerprint representations for practical molecule selection.

The use of learned fingerprints provides additional benefits across a number of tasks in the literature.[3,6,21] The benefits of using a learned embedding include limiting the over-valuation of biased features and fingerprint artifacts and accordingly may provide a better substrate to focus the learning algorithm. On the other hand, learned approaches incur a potential significant increase in computational cost both for the training phase and for the prediction phase, growing linearly with the size of the target prediction set without an option to pre-calculate the fingerprints and reuse them. Additionally, using learned fingerprints adds to the complexity of choosing the set of molecules to use for the embedding generation phase. While a self-referential approach, *i.e.* learning an embedding of the training data set itself, may be appropriate and provide additional means to focus on the most important chemical features, alternatives that may provide valuable insight would include fingerprints derived from the targeted prediction collection, or sets of bioactive and/or drug-like molecules, or combinations of such molecule sets.

Further, advanced and alternative molecular fingerprints may provide deeper information on the molecule. For example, pharmacophore-based fingerprints and 3D fingerprints that factor conformation[24,25] in addition to topological representation may provide a stronger basis for prediction of the interaction with protein binding pockets. Intriguingly for DNA-encoded library machine learning, a unique opportunity in representation presents itself where the fixed position of the DNA barcode provides a significant landmark for molecule orientation due to the physical requirement of solvent exposure of the DNA.

2.5 DEL-data Complexity

The justified application of warnings of 'garbage in, garbage out' is highly relevant in DEL-ML as with any machine learning application. With DEL, there are a number of complex interacting factors that may impact the quality of the inputs to machine learning training. Selection sequence data do not produce a high precision read-out of the binding that occurs in the DEL experiment. There is certainly stochastic 'bleed-through' of sequences that are driven by pure statistical chance. Such statistical processes in the resulting data are of significance with DEL data due to the substantial sequencing depth used in these experiments. Events that are 'normally' extremely rare may become relatively commonplace with 10–100s of millions of events (sequences) generated. However, perhaps more confounding is the capture of DEL molecules through non-productive binding events (*e.g.* promiscuous or non-specific binding), DNA-driven enhancement of modest binding, and the potential interplay between the molecular patterns of multiple protein conformations and binding sites. Where many traditional assays (*e.g.* those used in HTS) are driven by the mass action of many molecules to produce the readout, DELs are subject to the single molecular level precision of polymerase chain reaction (PCR). The benefit of this precision is that even very limited copy numbers of molecules in the input material can lead to detection events. On the flip side of this benefit, the presence of multiple alternate reaction-subpopulations of molecules for a given recorded reaction path in DEL synthesis may in fact be the source of the detected signal. Reaction products may include incomplete reaction intermediates and side products. Such alternate products are well-known in DEL and have increasingly been the subject of analysis through methods such ASMS[4,5] to improve the outcomes of traditional DEL analysis, *e.g.* in using the ASMS based testing with on-DNA synthesis to determine the true binding species. In machine learning applications however, such molecule-specific analysis would be impractical due to the scale, requiring acceptance that a DEL readout represents a range of possible molecules including truncates and unexpected side products. In fact, with a 50% cut-off in the yield ratio of the desired product by LCMS analysis on an example test reaction used by some DEL-synthesis practitioners, it is notable that, even provided such an yield ratio is consistent across a range of building blocks, the percentage of desired product across 3 cycles of such reactions would only produce the desired product 12.5% of the time. It is thus likely that, for many of the recorded DEL molecules, the 'expected' molecule represents a fraction of the species present for that barcode.

The complexity described above can be managed in multiple ways: through analysis approaches (*e.g.* denoising of data), through experimental approaches (*e.g.* repeats, multiple conditions), through molecular representation (*e.g.* providing multiple chemical representations for each encoding) and using the statistical power of machine learning itself to identify the complex chemical patterns in the data. A number of recent papers have discussed denoising of data[16,26] proposing a number of methods. In our work on DEL-ML[3] at X-Chem and Google Accelerated Sciences, we used a disynthon based approach to denoising data. Through aggregation of data across two cycle combinations of building blocks in larger libraries, the structures used for training must either show enrichment across a range of third cycle building blocks or to achieve a higher level of enrichment as a "tent-pole" molecule. The consistency of the signal generated by the two cycles limits the likelihood of spurious signal leading to over-representation or under-representation of the containing molecules. Given that the two-cycle building block combination population is more in line in terms of numbers with the sequencing depth, the approach reduces the likelihood of statistical anomaly driven classifications. Further, the aggregation plays out in both positive (*e.g.* target conditions) as well as negative selections (no target controls, competition selections) where the absence or presence of disynthon aggregated enrichment reduces the experimental variability in count-based signal on individual 3+ cycle molecules. The criticism of such approaches correctly points to the loss of information in the additional cycles, where the third cycle may not be constrained to a single building block, but nevertheless requires a specific range of chemistry or is enhanced differentially by different chemistry. On the one hand, this author believes that the overlapped nature of the disynthon information across the full selection does indeed contain much of that information. On the other, alternative approaches will likely provide valuable additional information to further specify and explore the desired chemical patterns. A hybrid two-step filtering approach may provide a strong compromise, where disynthon denoising provides an initial filter for full enumerated molecules which are prioritized further by counts. Other denoising approaches include enrichment metrics, statistical analyses,[3] and machine learning approaches[16] are used to improve the identification of molecules for resynthesis, which can equally be used as filters for machine learning training data points.

Additional data cleaning may be performed with selection condition-based filters. For example, a common approach in DEL data generation analysis includes the use of known competitive inhibitors with known binding modes. A filter, that includes or excludes the molecules that show an altered response in the presence of an inhibitor, 'cleans' the data by enhancing the desired patterns and reducing competing, undesired patterns. An interesting question arises in the treatment of these multiple conditions: does the removal of these data have the potential to negatively impact predictions? For example, in the case of a desired selectivity for a mutant isoform of a protein, removal of those binders to both wild-type and mutant protein may eliminate or even incorrectly train the model to avoid chemistry that is generally applicable in the binding pocket. An alternative approach of creating multiple models and assessing the presence or absence of predictions across these several models may provide a better strategy. Equally, multi-objective/multi-task learning paradigms may provide a mechanism for the inclusion of valuable information. Additional measures can improve data signal, including through performing repeats of the DEL experiment where the stochastic signal will fail to persist in a second selection in contrast to binding driven signal that will repeat. With promiscuous and non-specific binding elements, data cleaning can be enhanced using additional filters, *e.g.* of over-represented building blocks and the use of historic binding databases that cross over the large number of DEL experiments performed by an organization. Further, research efforts include not only the intended/encoded end product for model training but also the full range of side products and truncations.

2.6 Machine Learning Considerations

While machine learning creates new opportunities to identify complex patterns across extremely large data sets, it also creates a number of challenges. Foremost of these challenges is the potential for over-learning/memorization of the data and the learning of trivial patterns or the prediction of molecules that are impractical or cost-prohibitive to develop. As an example, DELs are limited in their synthesis to a subset of chemical reactions. This limitation impacts chemical diversity, chemical space coverage, and the property space of DEL molecules when compared to synthesizable 'drug-discovery' space. As a result, a concern with respect to DEL-ML is the learning of DEL-like chemical motifs that may artificially boost prediction signal and increase the likelihood of false positives. Inclusion of a broad diversity of DEL molecules in the negative data used for training is essential to reduce the overlearning of DEL-amenable chemistry. Further, inclusion of additional negative sources may provide a reduction in such signal. Under the assumption that (1) most molecules will be negatives against any particular protein and (2) a small number of false negatives will have limited impact, and the use of random molecules from reference sets (*e.g.* the targeted prediction collection, commercial or public databases of drug-like/bioactive molecules, *e.g.* ChEMBL, or molecules from other commercial or virtual DEL sources) may provide a valuable enhancement to the negative training signal.

Applying a post-prediction filter that combines prediction scores with negative predictive models based on one or more reference sets may improve the accuracy of DEL-ML.

In addition to refining negative data sets, DEL-ML benefits from assessment metrics and loss functions that more closely resemble the target problem. Unlike a diagnostic ML application where the goal is to identify accurately as much of a population with a condition without heavily misdiagnosing, DEL-ML is seeking to find a subset of candidate molecules as starting points within a vast chemical space where many chemical configurations may provide the required geometry. It is essential to focus DEL-ML on the reality of molecule assessment, which is heavily impacted by the time, money and resources that are required to assess molecules. In fact, for most practitioners of DEL and DEL-ML, the number of assessed molecules derived from an experiment will typically be no more than a few hundred. As a consequence, machine learning metrics for assessing the quality of trained models are less relevant when, as with receiver operator characteristic area-under the curve (ROC-AUC) and precision-recall area under the curve (PR-AUC) metrics typically used in the field, the metrics assess performance deep into the prediction-prioritized molecule list. In fact, the relevant assessment should be augmented by, if not replaced by, an analysis of early enrichment of active binders, *i.e.* hits in top percentile or top 1000 predictions. One approach to providing early enrichment information is the use of large-scale decoy molecule sets, *e.g.* from targeted prediction collections, and analyzing the rank order of predictions with respect to this reference set. While there may be false negatives in the decoy set, these are unlikely to represent a significant share. On the other hand, the positive set, whether training set positives or known internal or publicly disclosed positives, should show substantial early enrichment at the high end of the prediction score rankings. If the prediction of putative positives does not demonstrate significantly elevated prediction ranks compared to a reference set, then the model is unlikely to perform well in real-world tasks, *e.g.* hit-finding.

An additional challenge in the assessment of DEL-ML is the lack of diverse sets of 'ground-truth' positives in most applications. The challenge is further exacerbated by the fact that, as a result of the existence of a significant or even modest set of such molecules, the target in question is necessarily of lower interest. Consequently, splitting into training, testing and validation sets requires some thought in order to improve the likelihood of generalizability. As discussed in Yang *et al.*,[11] the use of different train/test/validate split factors can change the difficulty of the predictive task. Random splitting, while algorithmically simple, may present a relatively trivial task, particularly in the context of DELs where the overlap of molecular motifs is a result of the combinatorial synthetic approach. Alternative approaches for DEL data splitting include library-based splitting, library-disynthon based splitting, scaffold-based splitting, Tanimoto distance-based splitting, and combinations of these. While failure to generalize beyond the scope of the chemical space of the DEL may not negate the value of the model, the ability to generalize provides additional confidence that the learning achieved by the machine learning models extends beyond simple memorization or similarity-based pattern recognition.

A number of model types have been applied to DEL data, with the encouraging result that DEL data appear to be robust with respect to algorithm. Simple, tree-based models using algorithmically derived (as opposed to learned) fingerprint representations work well for identification of actives in published work.[3] While more complex models outperform in the published study, it is germane to note that the molecules discovered using these different models identify different chemical motifs. Accordingly, the combination of multiple methods may lead to improved outcomes in pursuing drug discovery projects. Given the limited prospective experimentation performed using DEL-ML methods, it remains to be seen to what degree algorithms impact the ability of models for example to (1) predict novel chemistry, (2) improve generalizability, (3) provide attributable prediction results, and (4) predict and prioritize within a series for optimization. Further, adoption of more complex algorithmic architectures may provide boosts in one or more of these areas. For example, the application of transfer learning, multi-task learning and multi-class learning may allow capitalization on large collections (*i.e.* covering many targets) of DEL results and improve the predictivity of models for targets with less strong responses to the initial DEL experiment. The use of regression models[26] has been demonstrated to provide improvements in some predictive tasks; however, due to the complexities of ground truth and validation data generation, these approaches have seen only limited exploration. In fact, it is plausible that, with the right set of DEL-conditions and treatment of the data, patterns that can be deconvoluted to binding affinity may exist in the data and may overcome the challenge of interpreting the high variability of DEL data and low fidelity of synthesis described above. The application of recent algorithmic approaches, *e.g.* diffusion, GANs, recurrent NNs, attention mechanisms, and others, may provide additional avenues for research.

2.7 Application of Models

Predictive models that have been generated have specific input requirements, for example,[27–29] groups have used HTS results as the required input information, where a subset of molecules may be tested against a specific protein of interest, and these results combined with broad results across decades of HTS results. In this extreme, the ability to analyze the 'next' molecule comes at a high cost. With DEL-ML, it is possible to use a variety of experimental information to generate the input data, including building block identities, conformers, and SMILES structures. In the case of SMILES structures, the input requirement for predicting any given molecule becomes trivial, thereby expanding the scope of prediction capability (not to be confused with applicability) to any molecule that can be expressed as a SMILES string and converted into a learned embedding or algorithmic fingerprint. Such a minimal requirement vastly expands the scope of what can be predicted and thereby opens up several opportunities that are more challenging when more complex computational (*e.g.* physics based calculations) or laboratory testing is required. As a result, DEL-ML studies have been able to capitalize on the rapidly broadening scope of accessible chemistry.[30] In fact, the minimal cost to apply prediction to the 'next' molecule enables the application of stricter filters for properties and attributes. One of the more interesting implications of such an approach to prediction inputs is that, in addition to drug-likeness, the cost of target molecules can be dialed in to allow vast reductions in the price of testing and simplification of reaction requirements (*i.e.* synthetic feasibility). Application of such filters can drive down the cost of experimentation and follow-up substantially by 10s to even 100s of fold reduction. With the rapid growth and scope of virtual collections such as Enamine REAL, Wuxi GalaXi, and Mcule Ultimate, the scope of low cost, accessible chemistry that can be readily predicted by the models has reached 10s to 100s of billions of molecules within the space of half a decade. Such models can also tap into the collections of pharma companies with large HTS sets creating the potential for reducing timelines of DEL selection to prediction to testing to a matter of days, limited only by intramural logistics. Some key considerations in the use of such collections should be factored in including the combinatorial or chemically redundant nature of these collections, which motivates the application of diversity filters in the prioritization of molecules for testing. Our work published on sEH in the Google/X-Chem DEL-ML study identified a heavy overrepresentation for adamantane containing molecules that were in the Mcule aggregated catalog. In addition to diversity considerations, commercial collections may contain significant percentages of molecules with toxicophores or poor properties—application of filters upfront will be valuable. Further, in the case of virtual collections, their size, which may soon reach into the trillions of molecules, motivates the pre-application of drug-likeness, toxicophore filters, and other filters in order to limit computational costs which even when relatively inexpensive on a per-molecule basis become prohibitive at the scale of 10^{12} or more. Additionally, pre-filtering of collections of molecules using best-in-class predictive models for ADME, toxicity, and PK properties holds potential for improving the developability of successful hits. Indeed, ML is driving parallel advances in these drug property models, and the combination with protein target-focused prediction provides a mechanism for improving the odds of downstream project success. Further, in assessing a broad virtual chemical space, careful consideration of the domain of applicability is warranted. While DEL-ML has been demonstrated to predict significantly outside of the DEL,[3] there may be significant limitations based on the specific DELs used, the scope of DEL chemistry available, and the algorithm used.

DEL-ML provides the potential for even further predictive value (1) through the downstream use of models validated in the initial hit-finding phase or (2) through the training of additional models enhanced by the results of the initial predictive experiments. In unpublished results, DEL-ML models have demonstrated an improved ability to predict molecules in chemical space proximal to the confirmed hits that were initially predicted by those models. Such predictive power around an initial hit creates the opportunity to triage follow-up molecules in medicinal chemistry optimization campaigns. With an effective model that predicts binding, medicinal chemists would be in a position to generate very large virtual libraries based on a seed or on multiple seed molecules. By providing prioritization through predictive scores, such models would allow greater confidence in multi-point modifications to the initial seed molecules with the potential to advance programs more rapidly and to reduce dead-ends in series exploration. Further, such prediction will not only enable structure–activity relationship (SAR) by catalog exploration, but also allow broader exploration of chemistry building on the existing virtual library synthesis infrastructure. For example, starting with chemical motifs that are validated in earlier rounds, the hit molecules can be functionalized at one or more positions with a reaction handle that enables the robust chemistry already in use in the virtual library. As a result, any building block with a compatible handle in the virtual library building block source can be used to expand that molecule. Each of these molecules can in turn be predicted and prioritized by the model. The use of large-scale virtual libraries for follow-up in this way can serve to substantially delay the much higher cost bespoke chemistry required to optimize a series while potentially providing a path to significantly advance chemistry and improve synthetic accessibility. As such programs advance and models are refined based on experimental results, the use of the DEL-ML models used in conjunction with multi-parameter optimizing models may allow medicinal chemists to prioritize molecules with a lower likelihood of dropping in potency while addressing liabilities and suboptimal properties, a trade-off that often plagues early drug discovery campaigns.

2.8 Applications of DEL-ML in the Literature

Since the publication of our DEL-ML study in 2020, a number of additional studies have been published or disclosed at conferences or in private. These studies have explored a growing number of DEL-ML elements. Machine learning has been reported to produce successful predictions across multiple sources of DNA encoded libraries, including both internal DEL collections as well as commercially available collections on libraries with significant ranges of diversity and size.[3,20] DEL-ML has been applied either prospectively or retrospectively[3,20,31–34] to a widening range of targets with companies reporting high rates of predictive success. Variations on models and approaches have demonstrated successful prediction using full molecule (rather than disynthon aggregated)[20] input data as well as the use of regression models.[26] Groups have reported successful prediction of toxicity end-points,[35] prediction using cell-based DEL data generation,[33] as well as the combined use of computational chemistry approaches with DEL-ML.[36] Additionally, DEL data derived ML has been demonstrated to be effective in the design of general and follow-up DELs.[18] The expansion of DEL-ML has now reached across a range of companies from small biotechs exploring a single target to mid-sized biotechs internalizing the capability to large pharma and AI-first discovery companies.

2.9 Challenges for DEL-ML

Despite initial success, significant hurdles remain for the widespread adoption of DEL-ML specifically and for predictive-AI/ML drug discovery approaches for small molecule therapeutics in general. Since our publication in 2020,[3] our study on the application of DEL-ML to 3 targets and testing of almost 2000 molecules remains relatively unique in the literature with respect to the scope of prospective testing. There are very limited publicly disclosed results and well-tested indications of best practices for DEL-ML and even less so for solving challenges posed by more difficult targets, more specific mechanisms of action, or selectivity profiles. Equally, little testing has been performed to improve the understanding of different algorithms, molecular representations, classification approaches, *etc.* This limitation of prospective data and publicly disclosed results extends beyond DEL-ML and has a chilling effect on the adoption of ML in small molecule drug discovery in general. Initial skepticism of machine learning often becomes a driver for confirmation bias with teams becoming convinced by underpowered studies. This is further exacerbated by the expectation that ML must not only equal the performance of traditional approaches, but vastly exceed the performance of these 'tried-and-true' methods that have been the subject of years of refinement and study and massive infusion of financial resources. Accordingly, underperformance in terms of the hit-rate may lead to a declaration of failure in a first experiment despite the cost being only a fraction of the cost of alternative methods. Failure to invest in the study of these promising technologies increases the potential that an effective but early approach, already demonstrating value in some cases, is relegated to the pile of underdeveloped technologies abandoned due to unreasonable expectations and underfunding. It is apparent at this moment in the industry that the infrastructure for small molecule predictive approaches is not where it ultimately needs to be in order to produce success. While teams declare that they can identify and test molecules in 2 to 3 months (already a time frame that is too slow for ML in drug discovery to have significant impact), the observable truth is that debate over experimental design, machine learning techniques, prediction result interpretation, and delays in synthesis and testing regularly stretch the time frame from initiation to assay results to twice as long, *i.e.* to the better part of a year. Such a time frame for ML based predictions will, correctly, 'lose' to more explicit medicinal chemistry design approaches. In the same 4-to-6-month time frame, a focused medicinal chemistry campaign can substantially advance even modest chemical material to improve potency and solve permeability, solubility, and initial ADME/Tox issues. Machine learning driven discovery can only deliver when it produces real-time, actionable predictions. Two key changes that are needed include (1) the development of rapid synthesis and testing approaches and (2) the shift from individual deterministic assessment of molecules to an information-oriented paradigm. ML is an information driven approach requiring significant data sets both for training and for experimental validation of models. Our infrastructure in the industry and our mindset are misaligned with these requirements—focused on individual results derived from full-scale synthesis and purification. A shift toward microscale and nanoscale approaches that produce just enough material, just-in-time and allowing for appropriate trade-offs is required. Building block acquisition, for example, represents 'dead-time' in the process and can be alleviated through prediction against only molecules that are easily made and have readily available starting material. Purification is notorious as a rate-limiting step. However, full-scale, HPLC purification may not be required for the production of valuable informer data and may be achievable through the use of scalable methods, *e.g.* filter plates. In addition, capitalizing on the predictive nature of DEL-ML and other ML in drug discovery approaches, it is possible to use data accumulated over time to learn what molecules can be readily produced with minimal likelihood of artifact producing impurities. The benefits of such rapid design–make–test–analyze (DMTA) cycles include the ability to better understand which models work best for a given data set, for expanding the breadth of chemical starting points, and for fuller exploration of series expansions. Without such a shift, the 'hype-cycle' default hypothesis of many of our colleagues will be a self-fulfilling prophesy and ML the easiest scapegoat for project failures.

2.10 Where Is the Field Headed?

DEL has steadily become a widely used hit-finding approach over the past few decades, *e.g.* ref. 37–43. Practitioners range from academics to big pharma and approaches have branched in a number of directions from the most common, DNA-recorded, solution phase approaches to DNA templated and solid-phase approaches. The benefits that DELs impart as an *in vitro* data generation source are poised to amplify the interest in these methods in the coming years provided the challenges of DEL-ML and of ML in drug discovery can be addressed. The current slow adaptation of predictive drug discovery can be overcome with the development of logistic and synthetic support that enables the rapid feedback loop required to support these new technologies.

With the favorable economics of DELs for generating data against a broad range of proteins and using a vast diversity of DEL-enabled chemistries, there is potential to generate a data set spanning the protein products of a significant portion of the genome. Figure 2.3 shows a future state where drug

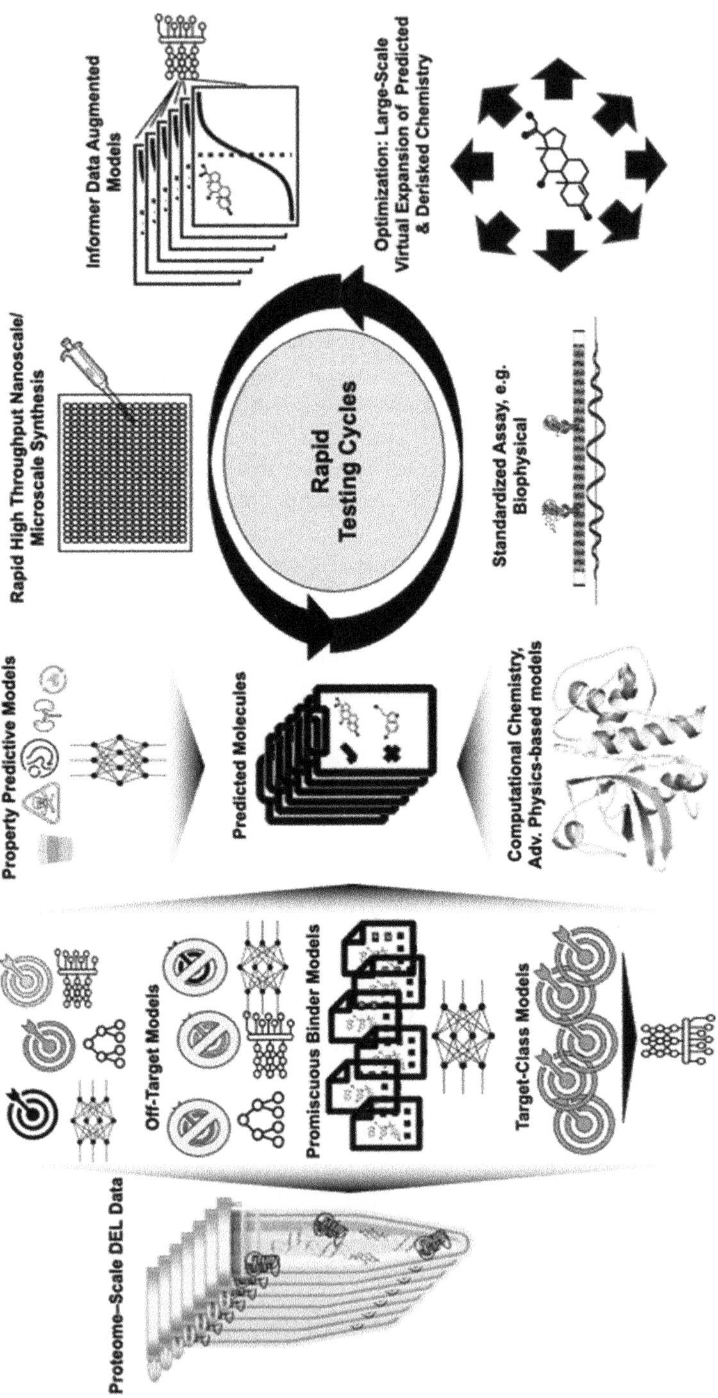
Proteome-Scale DEL Data
Target Construct Specific DEL-ML Models
Off-Target Models
Promiscuous Binder Models
Target-Class Models
Advanced ADME/Tox, Drug Property Predictive Models
Predicted Molecules
Computational Chemistry, Adv. Physics-based models
Rapid High Throughput Nanoscale/ Microscale Synthesis
Rapid Testing Cycles
Standardized Assay, e.g. Biophysical
Informer Data Augmented Models
Optimization: Large-Scale Virtual Expansion of Predicted & Derisked Chemistry

discovery can capitalize on this scale and scope of DEL data. In fact, several companies have already developed DEL data across hundreds and even thousands of proteins. Efforts have begun to capitalize on this approach, using both consortium and commercial means as well as scientific and automation-based cost reductions. The potential for such broad ranging chemical-biology data remains to be seen. A number of groups, *e.g.* ref. 44–46, have already demonstrated some degree of progress in prediction of small molecule-to-protein sequence relationships based on limited commercial, public and proprietary data. It is likely that data on a much larger scale are needed for a comprehensive predictive engine that achieves such a goal. DEL could be a key source for the scale of data needed for such efforts. Along the way to reaching such a 'moon-shot' result there are numerous benefits. The ability to comprehensively understand chemical-biology across an entire class of targets has shown value in the past decades (*e.g.* with kinases). Generating data across dozens to hundreds of proteins in a given class may serve to drive a similar understanding across lesser studied classes. The development of data sets addressing multiple proteins within a disease pathway will equally open opportunities to improve discovery outcomes where tool compounds may be more readily discovered for biological inquiry. In fact, these data sets and models have the potential to be used to identify molecules that affect more than one point of a cascade by focusing on the overlap of highly predicted molecules across multiple targets of interest.

Related is the potential for data at this scale to enable academic teams who have largely to date been financially limited from using DEL. First, the ability to provide large-scale DEL data to machine learning experts would

Figure 2.3 A possible future state of DEL-ML, achievable within years with the proper investment. The economics of DEL data generation is such that datasets can conceivably be created for large fractions of the disease relevant proteome. Generation of these data creates the opportunity for expanding the impact of DEL-ML. With a broad range of data sets, target and construct specific models can be generated. These models can be used both for target of interest prediction and for the avoidance of off-target interactions. The data sets across many targets can further serve as the basis for promiscuous binder identification models. In combination with data sets derived from targets of a similar class, models that identify privileged motifs for those classes may become applicable to a broader range of classes. Combined with advancing models for absorption, distribution, metabolism, excretion, toxicity (ADMET) and advancing physics-based methods, the molecule predictions will point to novel molecules with better properties and potency from the earliest stages. If combined with investment in rapid synthesis and testing of the predicted molecules at the micro- or nano-scale, rapid iteration will lead to volumes of informer data that can drive improved models. Similarly, the rapid iteration cycle can lead to an optimization path through prediction and testing of large-scale virtual libraries based on initial hits. This approach has the potential to both accelerate the process and improve the, currently poor, odds of success in drug development.

itself open opportunities in a community perhaps better equipped to try novel ML methods on the data. Second, the use of such data by groups to generate chemistry in the estimated 90%[47] of the genome products that have no tool molecules would provide unprecedented opportunities for academics to better understand complex disease biology. Lastly, combining data across multiple targets, multiple DEL sources, and across other non-DEL data sources is likely to bring improved results, *e.g.* by eliminating non-specific binders or enabling transfer learning applications.

The increasing capabilities of ML and computational methods are, of course, occurring in parallel in many other subfields. Combining physics-based approaches[36] with DEL-based methods has begun to be explored. As advanced approaches become increasingly affordable on a large scale and as novel methods are developed, the ability to use calculated and *in vitro* generated data together holds great potential. Of equal potential is the combination of DEL data and DEL-ML with steadily improving methods for multiparameter optimization, such as ADME/Tox, solubility, permeability, and PK. The simplest application pre-filters targeted molecule collections using property prediction models to limit to only those molecules with good drug properties before using the DEL-ML model for affinity prediction. More complex approaches would build models that simultaneously achieve multiple objectives of affinity and drug-likeness properties. Ultimately, the basis of the most successful AI/ML models to date has been the size, scope, and quality of the available data—natural language models, *e.g.* chat bots benefit from the vast sources of human generated written material. DEL offers a mechanism for providing data at the scale that may ultimately enable generative AI models that overcome challenges in the molecular predictive space and enable prediction for proteins and their isoforms without the need to perform the DEL-selection on that given protein construct.

In the interim, DEL-ML has begun to show the potential to move beyond DEL's more typical wheelhouse of initial hit-finding. DEL-ML models have been used in the driving of SAR by catalog and early potency optimization efforts. By providing a tool for increasing the odds of maintaining potency while addressing other molecule liabilities, activity/binding predictive models can already contribute to optimization campaigns. Beyond the iterative use of the model, DEL-ML can be augmented through the retraining of models based on the assessment of predicted molecules, such as by reweighting of data chemically similar to identified hits or through the factoring of experimentally derived potency. Fusing the use of DEL prediction with DEL synthesis itself presents an intriguing opportunity where the use of DEL-ML to predict the most informative downstream follow-up, focused-DEL represents an exciting workflow, particularly in cases where the synthesis of libraries is highly automated and building block availability is not a significant rate-limiting hurdle.

With increasing experimentation and interest in DEL-ML as an approach, the near-term goals in the field should include the broadening of target space, the identification of best practices, and the advancement of DEL-ML discovered molecules toward increasingly downstream discovery endpoints and ultimately to clinical trials and commercialization. Such successes will solidify the value of this approach, justify investment in the industry logistical changes that are needed, and create opportunity to explore the numerous avenues of research opened by the acceptance and broad utilization of molecular prediction on the proteome scale.

References

1. C. Nantasenamat, C. Isarankura-Na-Ayudhya, T. Naenna and V. Prachayasittikul, A practical overview of quantitative structure-activity relationship, *Excli J.*, 2009, **8**, 74–88.
2. S. Kong, R. Li, Y. Tian, Y. Zhang, Y. Lu, Q. Ou, P. Gao, K. Li and Y. Zhang, Single-cell omics: A new direction for functional genetic research in human diseases and animal models, *Front. Genet.*, 2023, **13**, 1100016.
3. K. McCloskey, E. A. Sigel, S. Kearnes, L. Xue, X. Tian, D. Moccia, D. Gikunju, S. Bazzaz, B. Chan, M. A. Clark, J. W. Cuozzo, M.-A. Guié, J. P. Guilinger, C. Huguet, C. D. Hupp, A. D. Keefe, C. J. Mulhern, Y. Zhang and P. Riley, Machine learning on DNA-encoded libraries: a new paradigm for hit finding, *J. Med. Chem.*, 2020, **63**, 8857–8866.
4. T. L. Foley, W. Burchett, Q. Chen, M. E. Flanagan, B. Kapinos, X. Li, J. I. Montgomery, A. S. Ratnayake, H. Zhu and M. C. Peakman, Selecting approaches for hit identification and increasing options by building the efficient discovery of actionable chemical matter from DNA-encoded libraries, *SLAS Discovery*, 2021, **26**, 263–280.
5. W. Su, R. Ge, D. Ding, W. Chen, W. Wang, H. Yan, W. Wang, Y. Yuan, H. Liu, M. Zhang and J. Zhang, Triaging of DNA-encoded library selection results by high-throughput resynthesis of DNA-conjugate and affinity selection mass spectrometry, *Bioconjugate Chem.*, 2021, **32**, 1001–1007.
6. S. Kearnes, K. McCloskey and M. Berndl, *et al.*, Molecular graph convolutions: moving beyond fingerprints, *J. Comput.-Aided Mol. Des.*, 2016, **30**, 595–608.
7. W. Chen, X. Liu, S. Zhang and S. Chen, Artificial intelligence for drug discovery: Resources, methods, and applications, *Mol. Ther.-Nucleic Acids*, 2023, **31**, 691–702.
8. Y. LeCun, Y. Bengio and G. Hinton, Deep learning, *Nature*, 2015, **521**, 436–444.
9. A. Krizhevsky, I. Sutskever and G. E. Hinton, ImageNet classification with deep convolutional neural networks, *Commun. ACM*, 2017, **60**, 84–90.
10. E. N. Muratov, J. Bajorath and R. P. Sheridan, *et al.*, QSAR without borders, *Chem. Soc. Rev.*, 2020, **49**, 3525–3564.
11. K. Yang, K. Swanson, W. Jin, C. Coley, P. Eiden, H. Gao, A. Guzman-Perez, T. Hopper, B. Kelley, M. Mathea, A. Palmer, V. Settels, T. Jaakkola, K. Jensen and R. Barzilay, Analyzing learned molecular representations for property prediction, *J. Chem. Inf. Model.*, 2019, **59**, 3370–3388.

12. V. G. Maltarollo, J. C. Gertrudes, P. R. Oliveira and K. M. Honorio, Applying machine learning techniques for ADME-Tox prediction: a review, *Expert Opin. Drug Metab. Toxicol.*, 2015, **11**, 259–271.
13. C. W. Coley, R. Barzilay, W. H. Green, T. S. Jaakola and K. F. Jensen, Convolutional Embedding of Attributed Molecular Graphs for Physical Property Prediction, *J. Chem. Inf. Model.*, 2017, **57**(8), 1757–1772.
14. J. Lyu, S. Wang and T. E. Balius, *et al.*, Ultra-large library docking for discovering new chemotypes, *Nature*, 2019, **566**, 224–229.
15. A. L. Satz, DNA encoded library selections and insights provided by computational simulations, *ACS Chem. Biol.*, 2015, **10**, 2237–2245.
16. P. Kómár and M. Kalinic, Denoising DNA encoded library screens with sparse learning, *ACS Comb. Sci.*, 2020, **22**, 410–421.
17. R. Hou, C. Xie, Y. Gui, G. Li and X. Li, Machine-learning-based data analysis method for cell-based selection of DNA-encoded libraries, *ACS Omega*, 2023, **8**, 19057–19071.
18. C. Zhang, M. Pitman, A. Dixit, S. Leelananda, H. Palacci, M. Lawler, S. Belyanskaya, L. Grady, J. Franklin, N. Tilmans and D. L. Mobley, Building block-based binding predictions for DNA-encoded libraries, *J. Chem. Inf. Model.*, 2023, **63**, 5120–5132.
19. H. Wang, G. Zhao, T. Zhang, Y. Li, G. Zhang and Y. Li, Comparative study of DNA barcode integrity evaluation approaches in the early-stage development of DNA-compatible chemical transformation, *ACS Pharmacol. Transl. Sci.*, 2023, **6**, 1724–1733.
20. W. Torng, I. Biancofiore, S. Oehler, J. Xu, J. Xu, I. Watson, B. Masina, L. Prati, N. Favalli, G. Bassi, D. Neri, S. Cazzamalli and J. A. Fen, Deep learning approach for the discovery of tumor-targeting small organic ligands from DNA-encoded chemical libraries, *ACS Omega*, 2023, **8**, 25090–25100.
21. R. Gómez-Bombarelli, J. N. Wei, D. Duvenaud, J. M. Hernández-Lobato, B. Sánchez-Lengeling, D. Sheberla, J. Aguilera-Iparraguirre, T. D. Hirzel, R. P. Adams and A. Aspuru-Guzik, Automatic chemical design using a data-driven continuous representation of molecules, *ACS Cent. Sci.*, 2018, **4**, 268–276.
22. RDKit: Open-source cheminformatics, https://www.rdkit.org.
23. Citadel Discovery Inc., 2023, Unpublished.
24. S. D. Axen, X. P. Huang, E. L. Cáceres, L. Gendelev, B. L. Roth and M. J. Keiser, A simple representation of three-dimensional molecular structure, *J. Med. Chem.*, 2017, **60**, 7393–7409.
25. Y. Wang, J. Hu, J. Lai, Y. Li, H. Jin, L. R. Zhang and Z. Liu, TF3P: three-dimensional force fields fingerprint learned by deep capsular network, *J. Chem. Inf. Model.*, 2020, **60**, 2754–2765.
26. K. S. Lim, A. G. Reidenbach, B. K. Hua, J. W. Mason, C. J. Gerry, P. A. Clemons and C. W. Coley, Machine learning on DNA-encoded library count data using an uncertainty-aware probabilistic loss function, *J. Chem. Inf. Model.*, 2022, **62**, 2316–2331.

27. S. Paricharak, A. P. IJzerman, A. Bender and F. Nigsch, Analysis of iterative screening with stepwise compound selection based on Novartis in-house HTS data, *ACS Chem. Biol.*, 2016, **11**, 1255–1264.
28. G. H. Dreiman, M. Bictash, P. V. Fish, L. Griffin and F. Svensson, Changing the HTS paradigm: AI-driven iterative screening for hit finding, *SLAS Discovery*, 2021, **26**, 257–262.
29. K. Y. Helal, M. Maciejewski, E. Gregori-Puigjané, M. Glick and A. M. Wassermann, Public Domain HTS Fingerprints: Design and Evaluation of Compound Bioactivity Profiles from PubChem's Bioassay Repository, *J. Chem. Inf. Model.*, 2016, **56**, 390–398.
30. M. Korn, C. Ehrt, F. Ruggiu, M. Gastreich and M. Rarey, Navigating large chemical spaces in early-phase drug discovery, *Curr. Opin. Struct. Biol.*, 2023, **80**, 102578.
31. 10th Annual DEL Symposium, Nov 3–4, 2022. Matthew Clark, X-Chem; Letian Kuai, Wuxi.
32. S. Ahmad, J. Xu, J. A. Feng, A. Hutchinson, H. Zeng, P. Ghiabi, A. Dong, P. A. Centrella, M. A. Clark, M.-A. Guié, J. P. Guilinger, A. D. Keefe, Y. Zhang, T. Cerruti, J. W. Cuozzo, M. von Rechenberg, A. Bolotokova, Y. Li, P. Loppnau, A. Seitova, Y.-Y. Li, V. Santhakumar, P. J. Brown, S. Ackloo and L. Halabelian, Discovery of a First-in-Class Small-Molecule Ligand for WDR91 Using DNA-Encoded Chemical Library Selection Followed by Machine Learning, *J. Med. Chem.*, 2023, **66**, 16051–16061.
33. R. Hou, C. Xie, Y. Gui, G. Li and X. Li, Machine-Learning-Based Data Analysis Method for Cell-Based Selection of DNA-Encoded Libraries, *ACS Omega*, 2023, **8**, 19057–19071.
34. A. S. M. Li, S. Kimani, B. Wilson, M. Noureldin, H. González-Álvarez, A. Mamai, L. Hoffer, J. P. Guilinger, Y. Zhang, M. von Rechenberg, J. S. Disch, C. J. Mulhern, B. L. Slakman, J. W. Cuozzo, A. Dong, G. Poda, M. Mohammed, P. Saraon, M. Mittal, P. Modh, V. Rathod, B. Patel, S. Ackloo, V. Santhakumar, M. M. Szewczyk, D. Barsyte-Lovejoy, C. H. Arrowsmith, R. Marcellus, M.-A. Guié, A. D. Keefe, P. J. Brown, L. Halabelian, R. Al-awar and M. Vedadi, Discovery of Nanomolar DCAF1 Small Molecule Ligands, *J. Med. Chem.*, 2023, **66**, 5041–5060.
35. V. Blay, X. Li, J. Gerlach, F. Urbina and S. Ekins, Combining DELs and machine learning for toxicology prediction, *Drug Discovery Today*, 2022, **27**, 103351.
36. K. Shmilovich, B. Chen, T. Karaletsos and M. M. Sultan, DEL-Dock: molecular docking-enabled modeling of DNA-encoded libraries, *J. Chem. Inf. Model.*, 2023, **63**, 2719–2727.
37. M. Leimbacher, Y. Zhang, L. Mannocci, M. Stravs, T. Geppert, J. Scheuermann, G. Schneider and D. Neri, Discovery of small-molecule interleukin-2 inhibitors from a DNA-encoded chemical library, *Chem. – Eur. J.*, 2012, **18**, 7729–7737.
38. J. W. Cuozzo, P. A. Centrella, D. Gikunju, S. Habeshian, C. D. Hupp, A. D. Keefe, E. A. Sigel, H. H. Soutter, H. A. Thomson, Y. Zhang and M. A. Clark, Discovery of a Potent BTK Inhibitor with a Novel Binding Mode by

Using Parallel Selections with a DNA-Encoded Chemical Library, *ChemBioChem*, 2017, **18**, 864–871.
39. M. A. Clark, R. A. Acharya and C. C. Arico-Muendel, *et al.*, Design, synthesis and selection of DNA-encoded small-molecule libraries, *Nat. Chem. Biol.*, 2009, **5**, 647–654.
40. H. Deng, H. O'Keefe and C. P. Davie, *et al.*, Discovery of highly potent and selective small molecule ADAMTS-5 inhibitors that inhibit human cartilage degradation *via* encoded library technology (ELT), *J. Med. Chem.*, 2012, **55**, 7061–7079.
41. P. A. Harris, S. B. Berger and J. U. Jeong, *et al.*, Discovery of a first-in-class receptor interacting protein 1 (RIP1) kinase specific clinical candidate (GSK2982772) for the treatment of inflammatory diseases, *J. Med. Chem.*, 2017, **60**, 1247–1261.
42. D.-Y. Wang, Y. Cao, L.-Y. Zheng, L.-D. Chen, X.-F. Chen, Z.-Y. Hong, Z.-Y. Zhu, X. Li and Y.-F. Chai, Target Identification of Kinase Inhibitor Alisertib (MLN8237) by Using DNA-Programmed Affinity Labeling, *Chem. – Eur. J.*, 2017, **23**, 10906–10914.
43. L. K. Petersen, A. B. Christensen, J. Andersen, C. G. Folkesson, O. Kristensen, C. Andersen, A. Alzu, F. A. Sløk, P. Blakskjær, D. Madsen, C. Azevedo, I. Micco and N. J. V. Hansen, Screening of DNA-Encoded Small Molecule Libraries inside a Living Cell, *J. Am. Chem. Soc.*, 2021, **143**, 2751–2756.
44. D. Karasev, B. Sobolev, A. Lagunin, D. Filimonov and V. Poroikov, Prediction of Protein-ligand Interaction Based on Sequence Similarity and Ligand Structural Features, *Int. J. Mol. Sci.*, 2020, **21**, 8152.
45. A. Chatterjee, R. Walters and Z. Shafi, *et al.*, Improving the generalizability of protein-ligand binding predictions with AI-Bind, *Nat. Commun.*, 2023, **14**, 1989.
46. Y. Zhang, S. Li, K. Meng and S. Sun, Machine Learning for Sequence and Structure-Based Protein–Ligand Interaction Prediction, *J. Chem. Inf. Model.*, 2024, **64**, 1456–1472.
47. T. Oprea, C. Bologa and S. Brunak, *et al.*, Unexplored therapeutic opportunities in the human genome, *Nat. Rev. Drug Discovery*, 2018, **17**, 317–332.

CHAPTER 3

Solid-phase DNA-encoded Library Technology

JUAN HU,[a] JOHN P. BURDICK[a] AND BRIAN M. PAEGEL*[a,b]

[a] Department of Pharmaceutical Sciences, University of California, Irvine, California 92697, USA; [b] Departments of Chemistry & Biomedical Engineering, University of California, Irvine, California 92697, USA
*Email: bpaegel@uci.edu

3.1 Introduction

DNA-encoded library (DEL) technology[1–3] has become a staple in the drug hunter's early lead identification toolbox, offering nearly limitless opportunities to explore chemical space[4,5] predominantly through the lens of affinity binding. With the increasing complexity of biological targets[6] and exponentially expanding sophistication of machine learning applications,[7] functional DEL screening technology has become highly desirable. In pursuit of this capability, our laboratory developed one-bead-one-compound DEL (OBOC-DEL) technology, which integrates DEL principles with the unique advantages of solid-phase synthesis and bead screening. Chief among those advantages is that each bead displays multiple copies of one library member that can be released from the bead and participate in conventional screening assays within a discrete volume. Here, we describe the solid-phase DEL platform, including library design and synthesis, microfluidic screening systems, relevant statistical analyses, and applications in affinity-, activity-, cell-, and permeability-based screening.

Drug Discovery Series No. 85
DNA-encoded Library Technology for Drug Discovery
Edited by Guansai Liu, Casey J. Krusemark and Jin Li

Published by the Royal Society of Chemistry, www.rsc.org

3.2 Solid-phase DEL Design

The OBOC approach, pioneered by Lam and colleagues, yields libraries of beads wherein an individual bead displays many copies of a unique chemical entity.[8] DNA encodes each synthetic step of the split-and-pool combinatorial library synthesis.[9] The functionalized bead scaffold—usually polymer synthesis resin—is compatible with numerous chemical reaction types and functional screening modalities.[10] Solid-phase synthesis resin beads are commercially available with various common functional groups for installing sites to be used in library member synthesis and DNA encoding. Library synthesis entails alternating between chemical building block (BB) coupling and enzymatic ligation of DNA oligonucleotides that encode a chemical transformation, similar to conventional "on-DNA" DEL construction.[1] However, OBOC-DEL beads display the library member and DNA-encoding tag as two branches of a bifunctional linker, rather than coupled directly to one another. Other key advantages of the OBOC-DEL approach include the potential for fully automated handling, easy removal of starting materials during synthesis, ready retrieval of library beads post screening by filtration, and the opportunity to decode and mass analyze individual DEL beads for rigorous library synthesis quality control (QC).[11]

An OBOC-DEL bead displays both the library member and the DNA-encoding tag *via* a custom, bifunctional linker. DNA-encoding sites are installed using the standard DEL "headpiece DNA" (HDNA), which is a PEG-linked double-stranded DNA displaying a 5′-phosphate, a 3′-CC dinucleotide overhang, and an internal amine functional group (Figure 3.1A).[1,11,12] The HDNA is acylated with 5-azidopentanoic acid and the azido HDNA is then clicked to the alkyne-displaying linker on resin *via* copper-catalyzed azide–alkyne cycloaddition (CuAAC). The linker also contains an Fmoc-protected primary amine for library member synthesis.[11] DEL-compatible reactions (*e.g.*, acylation, reductive alkylation, *etc.*)[13] are used to install BBs followed by enzymatic DNA oligonucleotide ligation. Photocleavable linkers (PC-linker) enable subsequent UV-dependent release of only the library member from the bead while the DNA tag remains attached (Figure 3.1B).[14,15] Up to $\sim 10^{10}$ copies of the library member per 10-μm bead can be liberated into the surrounding media to participate in the screening assay.[14] In one example OBOC-DEL, 192 amino acid BBs and 288 carboxylic acid BBs (55 296 total members) are linked through amide bond formation (Figure 3.1C).[16] BB selection cheminformatics for this library biased library content toward the rule of five (Ro5) druglikeness.[17] In another example 3-cycle antibacterial-inspired OBOC-DEL, 96 amino acids, 4 azido proline hubs, and 192 carboxylic acids (73 728 total members) were coupled *via* two amide formation reactions.[18] This DEL explored antibacterial-like chemical space, with library members possessing higher molecular weights, more hydrogen-bond donors/acceptors, and increased hydrophobicity (Figure 3.1D).

OBOC-DELs offer several advantages over conventional on-DNA DELs with respect to synthesis and screening. One unique advantage is the opportunity to perform rigorous QC post library synthesis. OBOC-DELs are prepared using a mixture of synthesis resin bead sizes, including screening beads (10 µm diameter) and QC beads (160 µm diameter). At the end of library synthesis, individual QC beads are harvested, PCR amplified to obtain DNA barcodes, then acid-cleaved and mass analyzed to check full library member synthesis *versus* its DNA code.[11] In contrast, on-DNA DELs are highly complex mixtures that are intractable for such QC using current analytical techniques. Solid-phase synthesis also enables reaction in organic solvents using a well-established catalog of solid-phase chemistries. Inspired by the dual-scale library QC approach, we developed dual-scale combinatorial synthesis "reaction rehearsal"[19] to assess both reaction synthetic yield on large-scale synthesis beads and DNA damage *via* qPCR on magnetic "crash test" microbeads functionalized with sacrificial DNA tags.[13] These critical studies lead to the identification of broad-scope and mild solid-phase amide formation and reductive amination conditions. Later refinement of these procedures yielded virtually innocent amide formation conditions suitable for preparing OBOC-DELs of natural product-inspired macrocyclic peptides requiring 5+ amide formation and amine deprotection cycles.[20] Finally, and most importantly, the bead serves as a vehicle for delivering many copies of one library member to a discrete volume, a fundamental requirement for screening.

3.3 Microfluidic OBOC-DEL Screening

High-throughput screening (HTS), the workhorse lead finding technology in early drug discovery, requires three elements: a source of chemical diversity, scalable distribution of the library to discrete volumes, and an assay of biological activity. The microtiter well plate is the vessel of choice for parsing chemical library members into separate wells for both storage and screening.[21] Robotic automation is the backbone of the HTS operation, with fluid handlers moving compounds and assays into wells and robotic arms moving plates between different operations (incubation readers, *etc.*). Microfluidics was long studied to miniaturize and further economize and parallelize HTS.[22–24] However, the so-called "world-to-chip" interface barrier was extreme—there was no scalable interface between a compound library held in microplates and a microfluidic analyzer.

The OBOC library approach offered a scalable world-to-chip interface solution to the HTS miniaturization conundrum. Conventional OBOC beads (~100–200 µm diameter) harbor 10–100 pmol of the library member. This is almost enough material to create stock solutions in the microliter volumes of microplates—indeed an entire platform based on this concept was explored.[25,26] Droplet microfluidics, which can generate picoliter-scale water-in-oil droplets, rapidly rose to prominence as a next-generation paradigm for assay miniaturization, and exquisite fluid handling architectures were devised to interface microplate compound libraries with such microfluidic analyses.[27,28] However, the intersection of solid-phase chemical synthesis

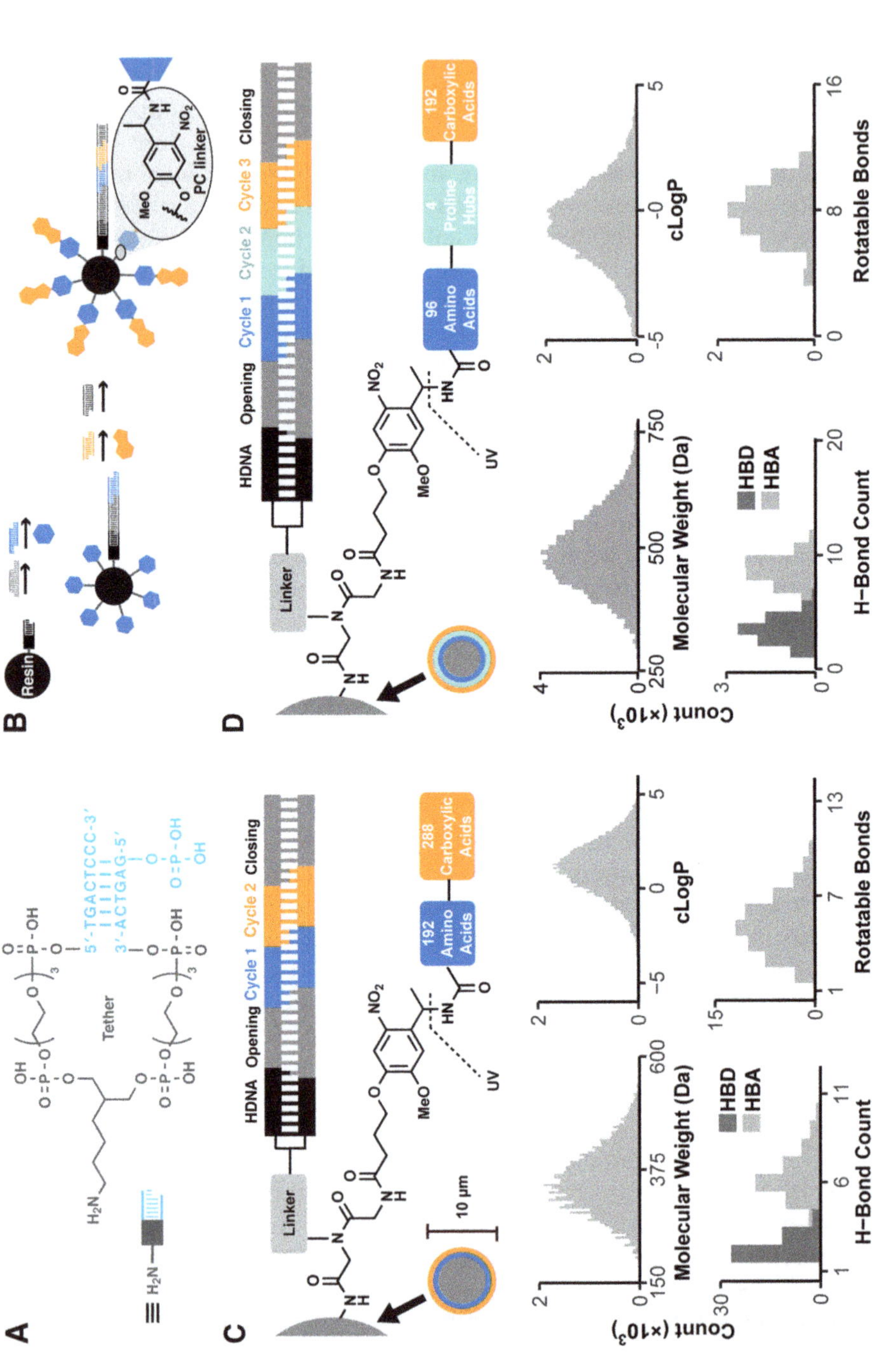
A
B
C
D
Resin
Tether
5'-TGACTCCC-3'
3'-ACTGAG-5'
PC linker
HDNA
Opening
Cycle 1
Cycle 2
Cycle 3
Closing
Linker
UV
10 μm
192 Amino Acids
288 Carboxylic Acids
96 Amino Acids
4 Proline Hubs
192 Carboxylic Acids
Count (×10³)
Molecular Weight (Da)
cLogP
H-Bond Count
Rotatable Bonds
HBD
HBA

and microfluidics offered compelling scaling advantages: depositing one of the smallest size synthesis resin beads (10 μm diameter, 100 fmol loading capacity) in a microfluidic water-in-oil droplet (~100 pL) generates ~1 mM concentrations of library member in the droplet were all of the bead's loading capacity to be released into the droplet. This back-of-the-envelope calculation and other foundational studies[29–31] set the stage for our early work exploring the feasibility of microfluidic miniaturization and automation toward OBOC-DEL functional screening.

Several engineering questions guided the first studies of microfluidic OBOC-DEL screening: (1) what is the most efficient method to load beads into droplets? (2) How can the library member be released from the bead? (3) Can all required steps be integrated into one device architecture? To load beads into the device, we developed a sedimentation-based approach called a "bead hopper." The approach exploits the bead's natural tendency to settle.[32] Photocleavage proved reliable for releasing the compound from beads. Our first study showed that bulk UV irradiation dose-dependently released the compound from beads and generated a 100 μM solution in the host droplet; pepstatin A photocleavage from

Figure 3.1 OBOC-DEL scheme and physicochemical property distributions. (A) The DNA "headpiece" is a PEG-tethered double-stranded DNA containing a dinucleotide overhang for enzymatic ligation and a primary amine. The amine is acylated with 5-azidopentanoic acid for subsequent coupling to the alkyne linker on library resin *via* CuAAC. (B) Encoded synthesis yields beads that display many copies of one DEL member type and its corresponding DNA encoding sequence. Library members are coupled to the solid support *via* an *o*-nitroveratryl photocleavable amino linker (PC linker). (C) All DEL tags begin with ligation of a common "opening" sequence and end with ligation of a common "closing" sequence. The opening and closing sequences are the primer binding sites for PCR amplification of encoding tags. A 2-cycle lead-like DEL was synthesized containing 55 296 members. In the first cycle, 192 Fmoc-protected amino acids were coupled and encoded by their corresponding DNA tag (blue). Following Fmoc deprotection, 288 carboxylic acids were coupled and encoded (orange). Library member physicochemical properties including molecular weight, partition coefficient (clog*P*), hydrogen bond donor and acceptor count (HBD, HBA), and rotatable bond count were calculated for a photocleaved primary amide DEL member (UV photocleavable bond indicated). (D) A 3-cycle antibacterial-like DEL was synthesized containing 73 728 members. In the first cycle, 96 amino acids were coupled and encoded by their corresponding DNA tag (blue). In the second cycle, 4 azido proline hubs were coupled and encoded (cyan). In the third cycle, 192 carboxylic acids were coupled and encoded (orange). The same library member properties were calculated for photocleaved primary amides. (A) Adapted from ref. 11 with permission from American Chemical Society, Copyright 2015. (B)–(D) Reproduced from ref. 12 and 16, https://doi.org/10.1021/acs.accounts.2c00791 and https://doi.org/10.1021/acsmedchemlett.3c00159, under the terms of the CC BY 4.0 license, https://creativecommons.org/licenses/by/4.0/.

beads robustly inhibited HIV protease activity in droplets. This assay was used to study integrated UV irradiation *via* onboard waveguide; HIV protease with photocleavable pepstatin A positive control "tool" beads was again a good testbed. Photocleavage in flow produced UV dose-dependent inhibition of HIV protease.[14]

OBOC-DEL microfluidic screening circuit integration continued to evolve toward fully automated bead handling, including loading, photodosing, incubation, and activity-based sorting. The integrated photodosing circuit was further elaborated with Frenz-type delay lines for integrated assay incubation[33] and a fluorescence-activated droplet sorting (FADS) component for isolating droplets containing bioactive beads.[34,35] The fully integrated circuit's performance was evaluated with photocleavable pepstatin A beads and a fluorogenic assay of cathepsin D activity. The device reliably sorted (99% accuracy) pepstatin beads from negative controls and a new encoding tag design was implemented to improve sequencing-based bead identification.[15,36] The Frenz delay lines, which feature numerous constrictions to redistribute droplets laterally during incubation, proved problematic with beads. A redesign of the incubator that relied on continuous oil phase extraction to close-pack the droplets and thereby defeat Poiseuille flow dispersion was developed. Oil extraction, or "emulsion creaming", is followed at the end of the incubation by reintroduction of oil to space and prepare droplets for sorting. This integrated continuous emulsion creamer ("ICE creamer") incubator figured prominently in later library screening circuit architectures.[37]

Our current working circuit architecture for OBOC-DEL activity-based screening integrates bead loading, compound photocleavage, droplet incubation, assay detection, and FADS. The circuit and these operations are schematized in Figure 3.2. OBOC-DEL beads with accompanying assay reagents are loaded into droplets at a rate of 1 bead per 10 droplets (1). Droplets are split twice using a Stone-type splitter[38] to increase throughput and reduce droplet volume (2). UV irradiation cleaves the DEL member from the bead surface into the encapsulating droplet (3). Photodosed droplets are incubated, allowing the biological activity assay to generate signal (4). Bioactive "hits" influence assay progression, causing the droplet fluorescence to deviate from the mean droplet fluorescence signal as measured in surrounding non-hit droplets. Laser-induced fluorescence enables high-speed, high-sensitivity detection of assay signal. Droplets that are statistically significantly deviated from the mean are dielectrophoretically sorted into a hit collection channel by applying a high voltage AC pulse (V_{AC}) to onboard electrodes (5). DNA tags of sorted beads are PCR amplified and sequenced en masse to deconvolute the BB composition of the sorted hits. This microfluidic OBOC-DEL analysis system provides the necessary architecture to facilitate an array of new biochemical screening assay modalities, and with it a variety of complex and desirable drug targets that are beyond the scope of traditional affinity selection.

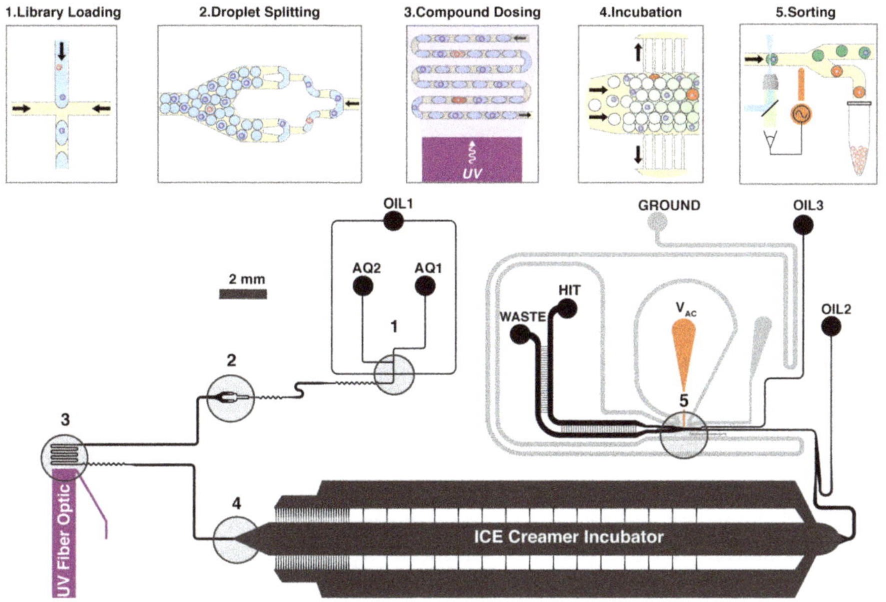

Figure 3.2 Integrated microfluidic screening architecture for functional DEL screening. OBOC-DEL beads enter the circuit together with the substrate (AQ1) and mix with the screening target (AQ2) immediately upstream of the water-in-oil (OIL1) flow-focusing junction (1), where droplets form (15 Hz). The flow continues toward the droplet splitter (2) to increase the throughput to 60 Hz. Compounds from the OBOC-DEL bead are photocleaved (365 nm) into the assay droplets as they traverse a serpentine photochemical reactor (3). Oil drains from the droplet flow, packing droplets into the integrated continuous emulsion creamer (ICE creamer) incubator (4) for assay incubation. At the incubator exit, two oil flows space (OIL2) and focus (OIL3) the droplet stream for analysis and sorting. Laser-induced fluorescence assay detection occurs upstream of the droplet sorting junction (5). Hit droplets that exhibit a statistically significantly different fluorescence compared to the overall droplet population trigger a high-voltage AC pulse to the working electrode (V_{AC}) that diverts hit droplets into the hit collection channel. Adapted from ref. 39 with permission from American Chemical Society, Copyright 2019.

3.4 Assay Development

Assay development is a core component of DEL screening that builds on decades of knowledge from the HTS field. For affinity DEL, assay development typically entails designing and expressing the protein target with subsequent rigorous biophysical characterization of target quality. OBOC-DEL assay development entails design and synthesis of components for functional assays (*e.g.*, target expression, synthesis of substrate "sensors", *etc.*). Assays transduce biological activity, such as enzyme inhibition or receptor binding, with production of detectable signal. Positive controls ("tool compounds") are used to optimize assay performance, which is quantified by measuring the difference in negative signal separation. The Z' score, a statistical parameter that measures mean positive and negative signal separation and their respective variance, quantitatively describes assay robustness. When $Z' = 0.5$, there are approximately 12 standard deviations (12σ) of separation between negative and positive control conditions and the assay is deemed suitable for HTS.[40,41] Performance must remain high through ultra-miniaturization of the assay to the picoliter droplet scale, which can sometimes face complications, such as assay component interactions with walls or reagent partitioning to the oil phase.[14,32,42]

Numerous fluorescence-based assays have been developed for microfluidic droplet OBOC-DEL screens. In each case, a flow injection analysis experiment is performed, wherein droplets are formed with all components of the activity assay (negative control), or in the presence of a known activity modulator (positive control). Normal distributions are fit to negative and positive control droplet fluorescence populations to obtain the mean and standard deviation values needed to calculate Z'.[14,32] A biochemical activity assay measures drug target autotaxin (ATX, a phosphodiesterase) enzyme activity *via* hydrolysis of a fluorogenic substrate (Figure 3.3A).[37,39] The inclusion of PF-8380 inhibits ATX activity and the fluorescence signal is statistically significantly attenuated compared to droplets with ATX and substrate ($Z' = 0.88$). A bacterial *in vitro* translation activity assay was adapted to microfluidic droplets to screen for inhibitors of bacterial protein synthesis as potential antibacterials (Figure 3.3B).[37] A DNA construct that encodes GFP along with bacterial IVTT reagents is encapsulated with and without the bacterial ribosomal protein synthesis inhibitor streptomycin, which significantly attenuates droplet fluorescence compared to droplets without streptomycin ($Z' = 0.64$). Two fluorescence polarization (FP) binding assays were also miniaturized to droplets using dye-conjugated ligands ("FP probe") of discoidin domain receptor 1 (DDR1) kinase and ATX (Figure 3.3C and D, respectively).[43] When bound to the macromolecular drug targets, the rotational rate of the probe's dipole is reduced, preserving fluorescence emission polarization.[44] Droplets containing either DDR1 or ATX and the respective FP probe were high FP signals compared to droplets with the probe alone ($Z' = 0.56$ and 0.67, respectively). Cumulatively, these studies established a robust, quantitative workflow for OBOC-DEL assay development.

3.5 OBOC-DEL Sampling and Screening Statistics

Combinatorial library synthesis and screening are fundamentally statistical exercises. Effective library design and accurate interpretation of screening data require an understanding of discrete probability distributions. Essential parameters, such as how many copies of a library member were likely sampled and screened ("representation"), what fraction of the theoretical library was sampled and screened ("coverage") as a function of

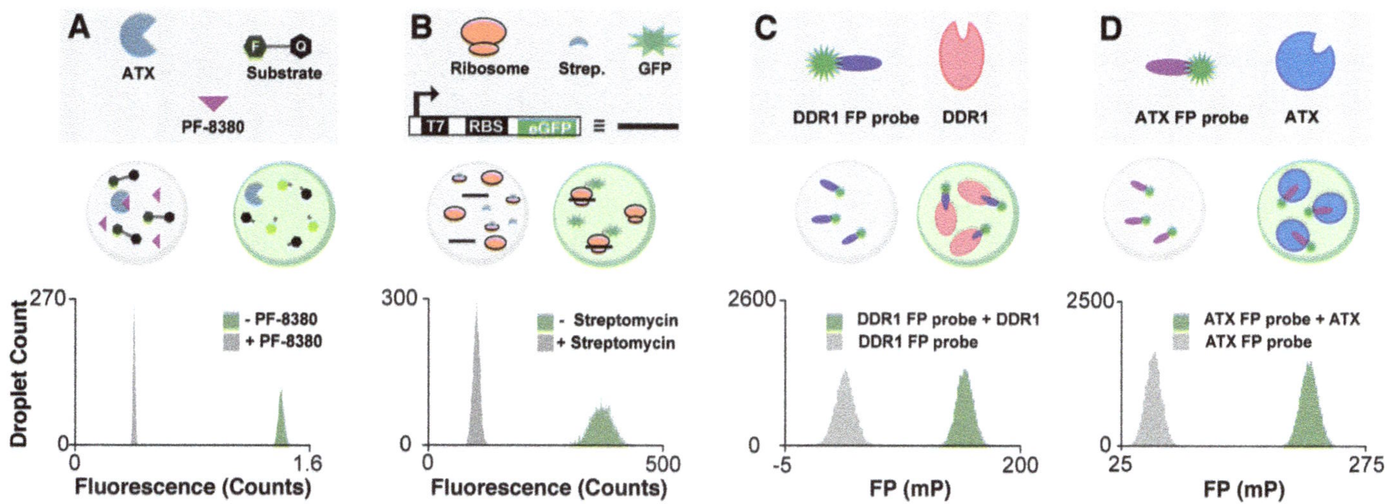

Figure 3.3 Droplet-scale assay development. For each assay, negative and positive control condition droplets are analyzed. The mean droplet laser-induced fluorescence and variance are extracted for each condition to calculate the droplet assay quality, Z' ($Z' > 0.5$ indicates suitability for HTS). (A) In a biochemical autotaxin (ATX) activity assay, droplets containing a fluorogenic substrate and enzyme are fluorescent; the positive control ATX inhibitor PF-8380 attenuates droplet fluorescence ($Z' = 0.88$). (B) In a bacterial translation activity assay, control droplets fluoresce upon GFP translation. Droplets containing the translation inhibitor streptomycin are low fluorescence ($Z' = 0.64$). In a fluorescence polarization (FP) assay of ligand binding, droplets with unbound fluorophore-labeled probes have low FP, while droplets with bound probes have high FP. Droplet-scale assay quality was suitable for both (C) DDR1 and (D) ATX binding assays ($Z' = 0.56$ and 0.67, respectively). Adapted from ref. 37, 39, and 43 with permission from American Chemical Society, Copyright 2017, 2019, and 2020, respectively.

representation, and occupancy of microfluidic droplets with beads, are all described using the Poisson distribution:

$$P(\lambda, k) = \frac{\lambda^k e^{-\lambda}}{k!} \tag{3.1}$$

where the probability, P, of observing a number of events occurring, k, is dependent on the mean expected number of events, λ. In applying this expression to combinatorial libraries, it is also useful to define a library equivalent, ε, as the number of library beads equal to the numerical diversity of the library (*i.e.*, 200 000 beads sampled from a 100 000-member library is 2ε).[45]

With library sampling formalism defined, we now apply the Poisson distribution to calculate the probability of encountering a given library member in different scenarios. First, let us take a random 200k-bead sample of a 100k-member library. Within this random 2ε sample, some library members will be present on one bead ($k=1$), some on 2 beads ($k=2$), and some members will not be present ($k=0$, Figure 3.4A).[46] The 2ε sample is a mean sampling of 2 ($\lambda=2$), and thus, for $k=(0, 1, 2, 3)$, we calculate $P=(0.14, 0.27, 0.27, 0.18)$; there is a 27% chance of observing any given member as a single bead, a 27% chance that a library member will be observed on 2 beads, and a 14% probability that a library member will not be present in the 2ε aliquot. The Poisson distribution therefore predicts that as the library sampling equivalents increase, so too does the representation of library members in higher k. From here, one can calculate the coverage of library members present at a given k class or higher using the Poisson cumulative distribution function. We performed extensive Monte Carlo simulations of ideal droplet sorting experiments, varying parameters such as the droplet occupancy with beads (λ_{drop}), hit rate (r), and equivalents screened (ε) with the goal of understanding the false discovery rate (FDR) dependence on the hit k class (Figure 3.4B). Given generous experimental conditions of 1 bead per droplet and a hit rate of 1.5%, Monte Carlo simulations indicated that restricting consideration of hits to those observed $k \geq 3$ beads results in FDR $<5\%$ while achieving nearly 100% library coverage (requiring screened $\sim 7\varepsilon$ of library). This theoretical framework catalyzed the development of bead-specific barcoding (BSB) informatics, wherein individual beads receive unique codes so that otherwise identical compound encoding regions can be differentiated during sequencing.[36]

Understanding the sources of and minimizing false discoveries is critical for encoded library screening. Encoded library screening hits must be synthesized and purified during validation, and thus high FDRs can rapidly erode the efficiency of combinatorial library synthesis and screening. Moreover, encoded library screening data sets are increasingly valuable as high-dimensionality data sets for training machine learning models.[7,47,48] The accuracy of model training is directly dependent on the FDR as the number of hits that can be validated as ground truth ($\sim 10^1$) is vastly lower

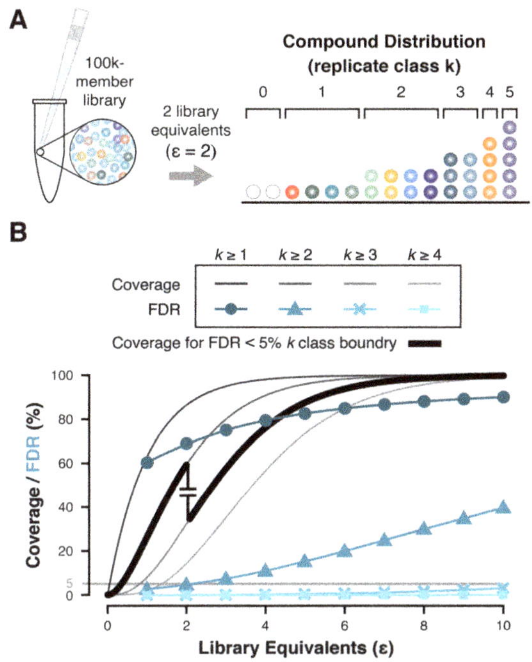

Figure 3.4 OBOC-DEL sampling and screening statistics. (A) A 2-equivalent (2ε) sample of a 100k-member DEL (200k beads) is used to predict library member representation. The Poisson distribution describes the probability of observing any given library member with replicate class *k*. (B) Fractional library coverage is plotted as a function of ε at each *k* class. The false discovery rate (FDR) is similarly plotted as a function of *k* and ε. One-hundred screening simulations, ranging from 1 to 10 library equivalents, were run for a 100k-member library with a 1.5% hit rate. Library coverage curves for each cumulative replicate class ($k \geq 1$–4, gray) are based on a Poisson cumulative distribution function. The FDR (%) was plotted as a function of ε for each cumulative replicate class ($k \geq 1$–4, teal). Coverage (bold trace) constrained by the FDR (<5%, gray horizontal) is calculated for each simulation iteration and the average is plotted (standard deviation markers are smaller than the data points). Simulation results are shown for average droplet occupancy $\lambda_{drop} = 1$. Adapted from ref. 46 with permission from American Chemical Society, Copyright 2017.

than the number of data points required for model building ($\sim 10^3$–10^6). The quality of OBOC-DEL functional screening data, which are particularly tantalizing for model training, depends on *r*, λ_{drop}, and ε screened. Screening sufficiently high ε is required to ensure that all library members are represented at sufficient *k* class, but as more ε are screened, the FDR of lower *k* classes starts to increase, requiring a higher *k* class cut-off to preserve acceptably low FDR. These simulations guided subsequent OBOC-DEL screening data interpretation[16,39,43,49] and will continue to prove invaluable with pending machine learning implementation. It is important to note that this theory only applies to random error, not systematic errors, such as those arising from, for example, pan-assay interference compounds (PAINS).[50,51]

3.6 OBOC-DEL Screening Examples

Identification of bioactive novel chemical matter from OBOC-DEL screening, which encompasses various competition binding, activity-based biochemical, and cellular modalities, integrates all previously described technical areas. Key early examples entailed the synthesis and screening of a 67 100-member Ro5-compliant 2-cycle OBOC-DEL. The library synthesis sequence comprised amino acid coupling followed by capping with a carboxylic acid. The first pilot screening target was the aforementioned phosphodiesterase ATX using the FS-3 fluorogenic substrate of ATX phosphodiesterase activity[52] and sorting droplets with statistically significantly lower fluorescence. Library screening identified numerous $k \geq 3$ hits with 20 out of 35 hits validating after synthesis.[39] In a second pilot screen of the same OBOC-DEL, ligands of the aforementioned DDR1 kinase domain were identified using an FP competition binding assay and sorting droplets with statistically significantly attenuated FP.[43] With these key successful demonstrations in hand, follow-on publications have disclosed the development of novel OBOC-DEL approaches that generate more granular structure–activity landscapes, such as dose-response and library-on-library screening strategies. Importantly, OBOC-DEL screening is enabling the investigation of various difficult targets and bioactivity, such as whole-cell bacterial cytotoxicity, RNA secondary structures, and membrane permeability using libraries of bespoke chemical matter.

3.6.1 Dose-response Activity-based OBOC-DEL Screening

As technology development efforts toward microfluidic activity-based OBOC-DEL screening progressed, it was clear that there were opportunities to layer the structure–activity data with compound dose by modulating the UV LED intensity. So-called "dose-response screening" or "quantitative HTS" (qHTS) had been shown to be highly effective for profiling chemical libraries because the IC_{50} analysis readily excluded compounds with non-conforming behavior.[53,54] Model OBOC-DEL screening experiments in microfluidic droplets demonstrated that model fluorescent compound photocleavage from droplets is dependent on UV intensity[32] and photocleavage of pepstatin A dose-dependently inhibited HIV protease activity in droplets.[14]

These studies were proof that on-line photocleavage and droplet assay detection could in principle recapitulate qHTS for OBOC-DEL screening. Dose-response OBOC-DEL screens used a 2-cycle 55 296-member druglike library to identify inhibitors of two model protein targets, the anticoagulant target protease factor Xa (FXa) and the previously studied phosphodiesterase ATX.[16] FXa inhibitor screens used a commercially available peptide substrate with the cleavage site flanked by a fluorophore (F) and quencher (Q) pair.

DEL screening occurred at multiple UV "doses" (LED intensities) monitoring hit rates at 4 or 6 standard deviations (4σ or 6σ) below the mean (Figure 3.5A–C), with an increased hit rate at a higher DEL member dose and a statistically insignificant hit rate with the LED off. Deconvolution of hits collected at full UV intensity and low dose revealed several high *k* class hit series and four notably overrepresented BBs (Figure 3.5D and E), including benzothienyl alanine **1**,

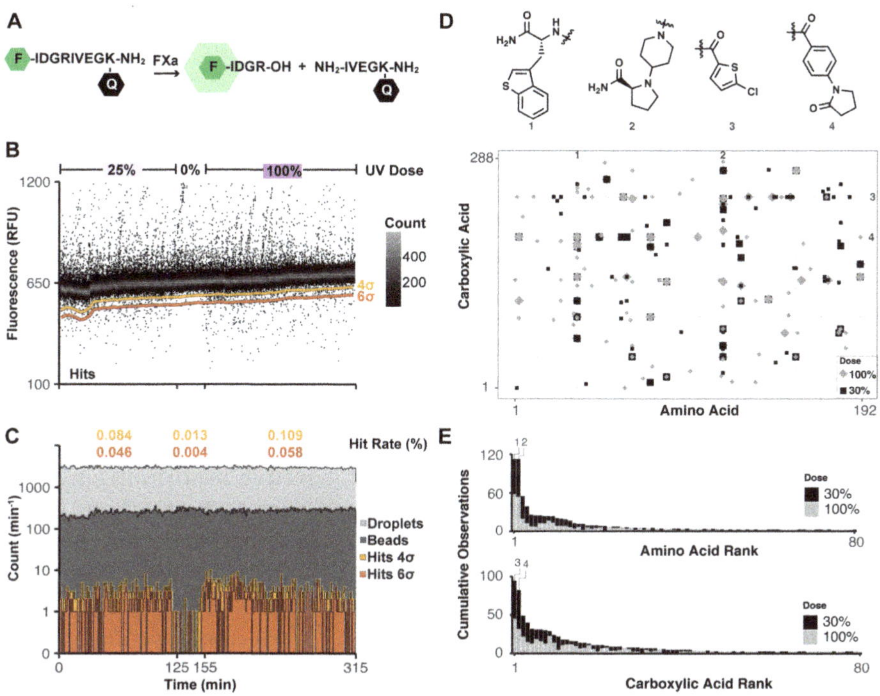

Figure 3.5 Dose-response activity-based OBOC-DEL screening for factor Xa (FXa). (A) FXa cleaves the substrate, a fluorophore quencher pair (F/Q) flanking a FXa peptide cleavage site, to liberate F from Q, increasing fluorescence. (B) Dose-response screenings were performed at three different UV intensities (100%, 0%, and 25% of maximum), yielding UV intensity-dependent hit rates. Fluorescence data were binned by time (1 min) and the dynamic sorting threshold (4 and 6 standard deviations, 4σ and 6σ, below the mean) is indicated (gold, orange). (C) Droplet, bead, and hit counts were binned in intervals (1 min) and plotted. (D) FXa dose-response DEL screening hits were sequenced to deduce hit structures. Data point shape and shading indicate UV intensity, and data point size indicates the hit replicate count (*k* class). Small points indicate $k = 2–3$ and large points indicate $k = 4–13$. Black squares represent 30% dosing and gray diamonds represent 100% dosing. (E) FXa amino acid and carboxylic acid representation was visualized for hits with $k > 1$. BB structures of the top 2 hits from each synthesis cycle are indicated. Reproduced from ref. 16, https://doi.org/10.1021/acsmedchemlett.3c00159, under the terms of the CC BY 4.0 license, https://creativecommons.org/licenses/by/4.0/.

piperidinyl proline 2, chlorothiophene 3, and phenylpyrrolidinone 4. Overrepresented BBs were hits at both high and low compound doses, suggesting conforming behavior, and 3/5 hits validated upon synthesis (including hits containing 3, a known FXa-binding fragment, and its isostere 1). ATX screening, hit deconvolution, and follow-up trends were similar: hits occurring at both UV doses tended to validate. Overall, dose-response screening adds a critical new layer of confidence to DEL screening data and could find further use in machine learning model building.

3.6.2 Cell-based Screening

Perhaps the most compelling application of OBOC-DEL functional screening is the ability to implement whole-cell assays to identify cell-active DEL members directly. A whole-cell bacterial cytotoxicity assay was the first demonstration of OBOC-DEL cell-based screening, owing to a robust literature precedent for OBOC screening for antibacterial compounds[55–57] and the important role DEL could play in exploring chemical space with a higher probability of antibacterial activity.[58–61] Combinatorial library beads were used in lawn-based bead diffusion assays to discover antimicrobial compounds.[55,57,62] Agar plates harboring either *E. coli* or *B. subtilis* were loaded with OBOC-DEL beads (Figure 3.6A). UV irradiation releases the library member from the bead into the agar. Following overnight culture, beads containing antibacterial compounds generate a growth inhibition zone (GIZ) and are extracted manually for subsequent DNA tag sequencing to decode the hit structures (Figure 3.6B and C). A 3-cycle OBOC-DEL was designed to explore chemical space biased toward physicochemical properties that correlate with antibacterials, such as higher molecular weight, lower clog*P*, and amine functionality (Figure 3.6D).[58,59] Hit compounds observed with high *k* class and containing overrepresented BBs (such as fluorochloronaphthyridone carboxylic acid) were synthesized for antibacterial testing. Hit synthesis uncovered an unanticipated S_NAr coupling to the aryl chloride, yielding whole families of structures with high similarity to ciprofloxacin. Of the compounds tested, 4/8 were active against *E. coli* or *B. subtilis* (Figure 3.6). Multi-drug resistant clinical isolates were unaffected, highlighting the need for further exploration.

3.6.3 On-bead Affinity-based Screening

The original envisioned application of OBOC-DEL[9,63] was binding-based assays on beads using fluorescence-activated cell sorting (FACS).[64] These initial disclosures occurred before the advent of genome sequencers, but they nonetheless recognized the powerful advantage of combining high-sensitivity PCR detection and structure elucidation *via* DNA sequencing to leverage FACS for rapid and automated bead screening. Our first example showcased an on-bead binding screen in the unbiased identification of "antigen surrogates," which specifically bound to immunoglobulin G (IgG)

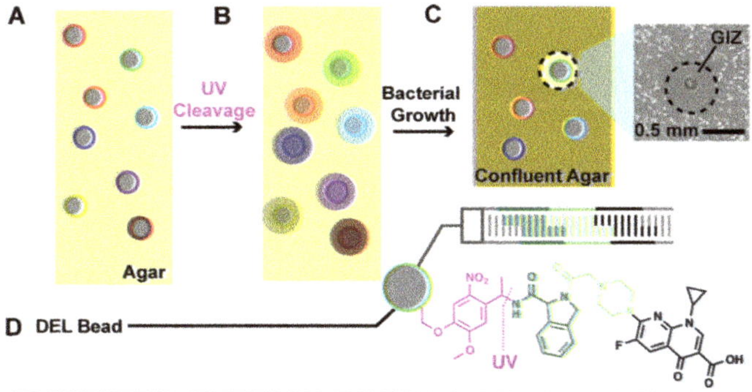

ID	Structure	*E. coli* MIC (μg/mL)	*B. subtilis* MIC (μg/mL)
1	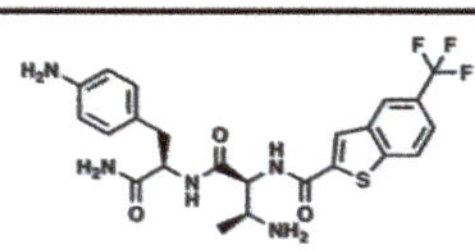	>32	1
2	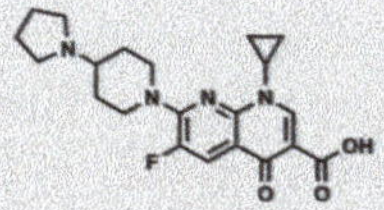	0.25	0.13
3	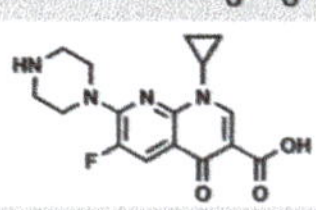	0.016	0.063
4	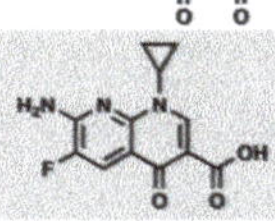	1	0.5
	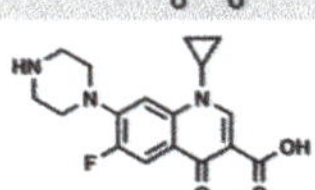 Ciprofloxacin	0.004	0.063

Figure 3.6 Antibacterial discovery *via* phenotypic DEL screening. (A) DEL beads are distributed on agar plates with bacteria. (B) UV irradiation of DEL beads liberates compounds from beads. The liberated compound diffuses through the surrounding agar. (C) A bacterial growth inhibition zone (GIZ) forms around hit beads after overnight incubation. (D) An antibacterial-like solid-phase DEL (3 cycles, 7488 members) was designed and screened against *E. coli* and *B. subtilis*. GIZ-inducing hit beads were sequenced to deduce hit structures. Unanticipated S_NAr chemistry yielded a series of elaborated ciprofloxacin hits (MIC reported in $\mu g\ mL^{-1}$). Adapted from ref. 49 with permission from American Chemical Society, Copyright 2021.

found in tuberculosis-infected patients.[36] In a 2-color screen, negative control patient IgGs were labeled with Alexa Flour 488 (AF488) anti-human IgG and active TB-infected (ATB) patient serum samples labeled with Alexa Flour 660 (AF660) anti-human IgG. The beads were mixed and screened together, gating on a low AF488/high AF660 population. Sorting and sequencing of these beads revealed a series of BBs that were statistically significantly enriched, including high *k* class oligomers that used all overrepresented BBs. These hits validated as highly selective ligands of ATB patient IgG. The experiment highlighted important design concepts for on-bead binding screens, the most important being the 2-color detection with target and negative control being as closely matched as possible (here the samples were both IgG populations), enabling stringent rejection of nonspecific binding.

These on-bead binding screen principles were then applied to the much more difficult problem of RNA ligand discovery. In an analogous 2-color screen, a tetramethyl rhodamine (TMR)-labeled hairpin RNA was mixed with a Dy647-labeled internal loop library (Dy647-ILL) containing 3×3 or 4096 (4^6) RNA secondary structures and exposed to a 73k-member DEL. Beads preferentially labeled with Dy647 were sequenced to reveal hit families. Several structures containing overrepresented BBs were synthesized and used to pull down the ILL sequences of that ligand *via* 2-dimensional combinatorial screening (2DCS).[65] These selections identified the products' target RNA, and subsequently a transcriptomic target candidate *via* "hit-StARTS,"[18] pri-miR-27a, a miRNA implicated in metastasis. These experiments also underscored an important role for DEL in discovering ligands of difficult targets by enabling a broad exploration of novel chemical matter *via* combinatorial synthesis. Disney's laboratory has continued to design, synthesize, and screen compound collections tailored to RNA ligand discovery, such as benzimidazole-focused libraries. Benzimidazoles are known to be excellent scaffolds for RNA ligands,[66] though the structure–activity relationships governing this selectivity are far from elucidated. In the study, ligands of the type I muscular dystrophy pathogenic RNA CUG expansion repeat that induces splicing defects *via* sequestration of muscleblind-like splicing regulator 1 were identified from on-bead OBOC-DEL screens.[67] The role of DELs in further prospecting RNA ligand chemical space continues to unfold.

3.6.4 Pharmacokinetic DEL Screening

The ability to prepare and screen novel and diverse chemical matter is a major advantage of DELs, but it also introduces substantial risk. Targets such as bacteria, RNA, or protein–protein interactions are likely to benefit from investigation using libraries of structures that violate the rules of druglikeness. Using hits from such screens as starting points for optimization can be risky if for no other reason than the parent structure's pharmacokinetic properties, which are almost always unknown, are likely to be poor depending on how severely the library's design violates the Ro5. Beyond computed properties, almost nothing is known about combinatorial library members other than their status as hits in screens or selections, and of those only a handful are ever destined for synthesis and characterization at scale. In contrast, HTS compound collections are exhaustively surveyed for pharmacokinetic properties, chief among them being membrane permeability, which correlates with oral bioavailability.[68] This observation spurred the deployment of the high-throughput parallel artificial membrane permeability assay (PAMPA) and cellular permeation assay technology.[69–71] High-throughput measurements of membrane permeability compatible with combinatorial libraries would critically inform design and hit prioritization for synthesis and follow-up.

Combinatorial library-scale pharmacokinetic screening capabilities required a suitably high-throughput membrane permeation assay and an accompanying method to detect all library members. A liposomal fluorogenic CuAAC reaction was developed to address both constraints.[72] Liposomes were loaded with a membrane-impermeable CalFluor 488 azide (CF-N_3) fluorogenic probe.[73] Membrane permeable alkyne-labeled molecules traverse the membrane and react with CF-N_3 to generate a fluorescent triazole product *via* CuAAC (Figure 3.7A). The alkyne was chosen as a minimal label that could be appended to all library members and not perturb the parent molecule's permeability.[74] The gold standard Z' analysis of assay quality was performed in microfluidic droplets of liposomal CF-N_3 using propargylamine (PPA) as the permeable positive control and methoxy-PEG17-COOH (mPEG17)

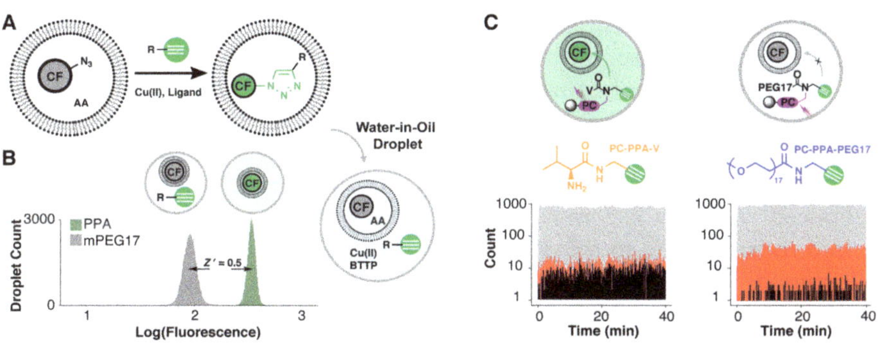

Figure 3.7 Permeation-based OBOC-DEL screening. (A) A liposomal permeation assay was devised wherein a membrane-impermeable fluorogenic azide probe (CF-N_3) is encapsulated with ascorbic acid (AA) in small unilamellar vesicles. Probe fluorescence increases only upon reaction with membrane-permeable alkyne (green) *via* CuAAC. (B) The high-throughput microfluidic droplet-scale permeation assay was conducted by encapsulating liposomal CF-N_3 and alkyne in droplets with Cu(II) and a ligand. Propargylamine (PPA) or mPEG17 was loaded into droplets together with liposomal CF-N_3. Droplets were incubated, and the CuAAC reaction progress was detected using laser-induced fluorescence. The assay quality score, Z', was calculated at $Z' = 0.5$ using PPA and mPEG17 as permeation positive and negative controls, respectively. (C) Control beads PC-PPA-V and PC-PPA-PEG17 were evaluated in droplet-scale liposomal permeation assays. The droplet generation rate (gray), bead occupancy rate (red), and hit rate (black) were binned for the PC-PPA-V and PC-PPA-PEG17 microfluidic analyses in 30 s windows. Adapted from ref. 72, https://doi.org/10.1021/acs.jmedchem.3c00138, under the terms of the CC BY 4.0 license, https://creativecommons.org/licenses/by/4.0/.

as the impermeable negative control, yielding $Z' = 0.5$ (Figure 3.7B). Proof-of-concept photocleavable control beads were prepared by coupling PPA to resin *via* reductive alkylation as a general alkyne label, then acylation with either a membrane-permeable valine-*N*-propargyl amide (PC-PPA-V) or membrane-impermeable PEG17 (PC-PPA-PEG17) and evaluated in microfluidic droplets with on-line photocleavage; nearly all droplets with PC-PPA-V were identified as hits, while PC-PPA-PEG17 exhibited minimal assay signal (Figure 3.7C), setting the stage for scalable pharmacokinetic OBOC-DEL screening.

3.7 Conclusions and Future Directions

Combining the principles of DEL and solid-phase synthesis has unlocked the ability to screen directly for function. Hits from activity-based screens tend to validate as active and they tend to be more tractable: 2-cycle ATX inhibitor OBOC-DEL screening hits were decidedly smaller than affinity hits, including several structures that did not survive affinity selection.[39] For applications like biochemical screening, these advantages may be modest in light of the much lower numerical diversity of OBOC-DELs and requirement of complex microanalytical instrumentation. For other applications, activity-based screening is essential. For example, whole-cell cytotoxicity screening is necessary for effective antibacterial discovery.[49] Eukaryotic cellular screening, another emergent capability, is important for granting DEL access to targets that cannot be expressed and purified (*e.g.*, complexes), are disordered, or only function in the context of cellular metabolism such as signal transduction cascades or protein degradation. Intracellular and proximity-enhanced selection is removing these obstacles[75–77] and OBOC-DEL is also poised to contribute to these efforts by recapitulating phenotypic cell-based HTS capabilities. Finally, assay development remains a major bottleneck for any screening modality. Scaling DEL and probe discovery more broadly to the human proteome demands a more plug-and-play strategy that is agnostic of protein structure or function. Efforts are afoot to realize such assay technology and in so doing begin to expand the frontiers of druggability.[12]

Acknowledgements

This work was supported by a grant award from the National Institutes of Health to B. M. P. (GM140890).

References

1. M. A. Clark, R. A. Acharya, C. C. Arico-Muendel, S. L. Belyanskaya, D. R. Benjamin, N. R. Carlson, P. A. Centrella, C. H. Chiu, S. P. Creaser, J. W. Cuozzo, C. P. Davie, Y. Ding, G. J. Franklin, K. D. Franzen, M. L. Gefter, S. P. Hale, N. J. V. Hansen, D. I. Israel, J. Jiang, M. J. Kavarana, M. S. Kelley, C. S. Kollmann, F. Li, K. Lind, S. Mataruse, P. F. Medeiros, J. A. Messer, P. Myers, H. O'Keefe, M. C. Oliff, C. E. Rise, A. L. Satz,

S. R. Skinner, J. L. Svendsen, L. Tang, K. van Vloten, R. W. Wagner, G. Yao, B. Zhao and B. A. Morgan, *Nat. Chem. Biol.*, 2009, **5**, 647–654.
2. R. A. Goodnow, C. E. Dumelin and A. D. Keefe, *Nat. Rev. Drug Discovery*, 2016, **16**, 131–147.
3. A. L. Satz, A. Brunschweiger, M. E. Flanagan, A. Gloger, N. J. V. Hansen, L. Kuai, V. B. K. Kunig, X. Lu, D. Madsen, L. A. Marcaurelle, C. Mulrooney, G. O'Donovan, S. Sakata and J. Scheuermann, *Nat. Rev. Methods Primers*, 2022, **2**, 1–17.
4. A. L. Satz, *ACS Chem. Biol.*, 2015, **10**, 2237–2245.
5. P. R. Fitzgerald and B. M. Paegel, *Chem. Rev.*, 2020, **12**, 7155–7177.
6. Y. Huang, Y. Li and X. Li, *Nat. Chem.*, 2022, **14**, 129–140.
7. K. McCloskey, E. A. Sigel, S. Kearnes, L. Xue, X. Tian, D. Moccia, D. Gikunju, S. Bazzaz, B. Chan, M. A. Clark, J. W. Cuozzo, M.-A. Guié, J. P. Guilinger, C. Huguet, C. D. Hupp, A. D. Keefe, C. J. Mulhern, Y. Zhang and P. Riley, *J. Med. Chem.*, 2020, **63**, 8857–8866.
8. K. S. Lam, S. E. Salmon, E. M. Hersh, V. J. Hruby, W. M. Kazmierski and R. J. Knapp, *Nature*, 1991, **354**, 82–84.
9. S. Brenner and R. A. Lerner, *Proc. Natl. Acad. Sci. U. S. A.*, 1992, **89**, 5381–5383.
10. K. S. Lam, M. Lebl and V. Krchňák, *Chem. Rev.*, 1997, **97**, 411–448.
11. A. B. MacConnell, P. J. McEnaney, V. J. Cavett and B. M. Paegel, *ACS Comb. Sci.*, 2015, **17**, 518–534.
12. A. Dixit, H. Barhoosh and B. M. Paegel, *Acc. Chem. Res.*, 2023, **56**, 489–499.
13. M. L. Malone and B. M. Paegel, *ACS Comb. Sci.*, 2016, **18**, 182–187.
14. A. K. Price, A. B. MacConnell and B. M. Paegel, *Anal. Chem.*, 2016, **88**, 2904–2911.
15. A. B. MacConnell, A. K. Price and B. M. Paegel, *ACS Comb. Sci.*, 2017, **19**, 181–192.
16. P. R. Fitzgerald, W. G. Cochrane and B. M. Paegel, *ACS Med. Chem. Lett.*, 2023, **14**, 1295–1303.
17. C. A. Lipinski, F. Lombardo, B. W. Dominy and P. J. Feeney, *Adv. Drug Delivery Rev.*, 2001, **46**, 3–26.
18. R. I. Benhamou, B. M. Suresh, Y. Tong, W. G. Cochrane, V. Cavett, S. Vezina-Dawod, D. Abegg, J. L. Childs-Disney, A. Adibekian, B. M. Paegel and M. D. Disney, *Proc. Natl. Acad. Sci. U. S. A.*, 2022, **119**, e2114971119.
19. H. M. Geysen, C. D. Wagner, W. M. Bodnar, C. J. Markworth, G. J. Parke, F. J. Schoenen, D. S. Wagner and D. S. Kinder, *Chem. Biol.*, 1996, **3**, 679–688.
20. A. K. Price and B. M. Paegel, *ACS Comb. Sci.*, 2020, **22**, 649–655.
21. D. S. Auld, P. A. Coassin, N. P. Coussens, P. Hensley, C. Klumpp-Thomas, S. Michael, G. S. Sittampalam, O. J. Trask, B. K. Wagner, J. R. Weidner, M. J. Wildey and J. L. Dahlin, in *Assay Guidance Manual, Eli Lilly & Company and the National Center for Advancing Translational Sciences*, Bethesda (MD), 2004.
22. B. L. Wang, A. Ghaderi, H. Zhou, J. Agresti, D. A. Weitz, G. R. Fink and G. Stephanopoulos, *Nat. Biotechnol.*, 2014, **32**, 473–478.

23. R. Ostafe, R. Prodanovic, W. Lloyd Ung, D. A. Weitz and R. Fischer, *Biomicrofluidics*, 2014, **8**, 041102.
24. G. Woronoff, M. Ryckelynck, J. Wessel, O. Schicke, A. D. Griffiths and P. Soumillion, *ChemBioChem*, 2015, **16**, 1343–1349.
25. R. Liang, L. Yan, J. Loebach, M. Ge, Y. Uozumi, K. Sekanina, N. Horan, J. Gildersleeve, C. Thompson, A. Smith, K. Biswas, W. C. Still and D. Kahne, *Science*, 1996, **274**, 1520–1522.
26. H. E. Blackwell, L. Pérez, R. A. Stavenger, J. A. Tallarico, E. Cope Eatough, M. A. Foley and S. L. Schreiber, *Chem. Biol.*, 2001, **8**, 1167–1182.
27. S. Sun and R. T. Kennedy, *Anal. Chem.*, 2014, **86**, 9309–9314.
28. E. D. Guetschow, D. J. Steyer and R. T. Kennedy, *Anal. Chem.*, 2014, **86**, 10373–10379.
29. A. J. You, R. J. Jackman, G. M. Whitesides and S. L. Schreiber, *Chem. Biol.*, 1997, **4**, 969–975.
30. W. H. Tan and S. Takeuchi, *Proc. Natl. Acad. Sci. U. S. A.*, 2007, **104**, 1146–1151.
31. G. Upert, C. A. Merten and H. Wennemers, *Chem. Commun.*, 2010, **46**, 2209.
32. A. K. Price, A. B. MacConnell and B. M. Paegel, *Anal. Chem.*, 2014, **86**, 5039–5044.
33. L. Frenz, K. Blank, E. Brouzes and A. D. Griffiths, *Lab Chip*, 2009, **9**, 1344–1348.
34. K. Ahn, C. Kerbage, T. P. Hunt, R. M. Westervelt, D. R. Link and D. A. Weitz, *Appl. Phys. Lett.*, 2006, **88**, 024104.
35. A. Sciambi and A. R. Abate, *Lab Chip*, 2015, **15**, 47–51.
36. K. R. Mendes, M. L. Malone, J. M. Ndungu, I. Suponitsky-Kroyter, V. J. Cavett, P. J. McEnaney, A. B. MacConnell, T. M. Doran, K. Ronacher, K. Stanley, O. Utset, G. Walzl, B. M. Paegel and T. Kodadek, *ACS Chem. Biol.*, 2016, **12**, 234–243.
37. W. G. Cochrane, A. L. Hackler, V. J. Cavett, A. K. Price and B. M. Paegel, *Anal. Chem.*, 2017, **89**, 13227–13234.
38. D. R. Link, S. L. Anna, D. A. Weitz and H. A. Stone, *Phys. Rev. Lett.*, 2004, **92**, 054503.
39. W. G. Cochrane, M. L. Malone, V. Q. Dang, V. Cavett, A. L. Satz and B. M. Paegel, *ACS Comb. Sci.*, 2019, **21**, 425–435.
40. J. H. Zhang, T. D. Y. Chung and K. R. Oldenburg, *J. Biomol. Screening*, 1999, **4**, 67–73.
41. *Assay Guidance Manual*, ed. S. Markossian, A. Grossman, M. Arkin, D. Auld, C. Austin, J. Baell, K. Brimacombe, T. D. Y. Chung, N. P. Coussens, J. L. Dahlin, V. Devanarayan, T. L. Foley, M. Glicksman, K. Gorshkov, J. V. Haas, M. D. Hall, S. Hoare, J. Inglese, P. W. Iversen, M. Lal-Nag, Z. Li, J. R. Manro, J. McGee, O. McManus, M. Pearson, T. Riss, P. Saradjian, G. S. Sittampalam, M. Tarselli, O. J. Trask, J. R. Weidner, M. J. Wildey, K. Wilson, M. Xia and X. Xu, Eli Lilly & Company and the National Center for Advancing Translational Sciences, Bethesda (MD), 2004.

42. F. Courtois, L. F. Olguin, G. Whyte, A. B. Theberge, W. T. S. Huck, F. Hollfelder and C. Abell, *Anal. Chem.*, 2009, **81**, 3008–3016.
43. A. L. Hackler, F. G. FitzGerald, V. Q. Dang, A. L. Satz and B. M. Paegel, *ACS Comb. Sci.*, 2019, **22**, 25–34.
44. A. M. Rossi and C. W. Taylor, *Nat. Protoc.*, 2011, **6**, 365–387.
45. J. J. Burbaum, M. H. Ohlmeyer, J. C. Reader, I. Henderson, L. W. Dillard, G. Li, T. L. Randle, N. H. Sigal, D. Chelsky and J. J. Baldwin, *Proc. Natl. Acad. Sci. U. S. A.*, 1995, **92**, 6027–6031.
46. A. B. MacConnell, A. K. Price and B. M. Paegel, *ACS Comb. Sci.*, 2017, **19**, 524–532.
47. K. S. Lim, A. G. Reidenbach, B. K. Hua, J. W. Mason, C. J. Gerry, P. A. Clemons and C. W. Coley, *J. Chem. Inf. Model.*, 2022, **62**, 2316–2331.
48. R. Hou, C. Xie, Y. Gui, G. Li and X. Li, *ACS Omega*, 2023, **8**, 19057–19071.
49. W. G. Cochrane, P. R. Fitzgerald and B. M. Paegel, *ACS Chem. Biol.*, 2021, **16**, 2752–2756.
50. J. B. Baell and G. A. Holloway, *J. Med. Chem.*, 2010, **53**, 2719–2740.
51. J. B. Baell and J. W. M. Nissink, *ACS Chem. Biol.*, 2017, **13**, 36–44.
52. C. G. Ferguson, C. S. Bigman, R. D. Richardson, L. A. van Meeteren, W. H. Moolenaar and G. D. Prestwich, *Org. Lett.*, 2006, **8**, 2023–2026.
53. J. Inglese, D. S. Auld, A. Jadhav, R. L. Johnson, A. Simeonov, A. Yasgar, W. Zheng and C. P. Austin, *Proc. Natl. Acad. Sci. U. S. A.*, 2006, **103**, 11473–11478.
54. O. J. Miller, A. E. Harrak, T. Mangeat, J. C. Baret, L. Frenz, B. E. Debs, E. Mayot, M. L. Samuels, E. K. Rooney, P. Dieu, M. Galvan, D. R. Link and A. D. Griffiths, *Proc. Natl. Acad. Sci. U. S. A.*, 2011, **109**, 378–383.
55. J. L. Silen, A. T. Lu, D. W. Solas, M. A. Gore, D. Maclean, N. H. Shah, J. M. Coffin, N. S. Bhinderwala, Y. Wang, K. T. Tsutsui, G. C. Look, D. A. Campbell, R. L. Hale, M. Navre and C. R. DeLuca-Flaherty, *Antimicrob. Agents Chemother.*, 1998, **42**, 1447–1453.
56. V. S. Fluxà, N. Maillard, M. G. P. Page and J. L. Reymond, *Chem. Commun.*, 2011, **47**, 1434–1436.
57. K. J. Fisher, J. A. Turkett, A. E. Corson and K. L. Bicker, *ACS Comb. Sci.*, 2016, **18**, 287–291.
58. D. J. Payne, M. N. Gwynn, D. J. Holmes and D. L. Pompliano, *Nat. Rev. Drug Discovery*, 2007, **6**, 29–40.
59. M. F. Richter, B. S. Drown, A. P. Riley, A. Garcia, T. Shirai, R. L. Svec and P. J. Hergenrother, *Nature*, 2017, **545**, 299–304.
60. J. M. Stokes, K. Yang, K. Swanson, W. Jin, A. Cubillos-Ruiz, N. M. Donghia, C. R. MacNair, S. French, L. A. Carfrae, Z. Bloom-Ackermann, V. M. Tran, A. Chiappino-Pepe, A. H. Badran, I. W. Andrews, E. J. Chory, G. M. Church, E. D. Brown, T. S. Jaakkola, R. Barzilay and J. J. Collins, *Cell*, 2020, **180**, 688–702.e13.
61. C. Zampaloni, P. Mattei, K. Bleicher, L. Winther, C. Thäte, C. Bucher, J.-M. Adam, A. Alanine, K. E. Amrein, V. Baidin, C. Bieniossek, C. Bissantz, F. Boess, C. Cantrill, T. Clairfeuille, F. Dey, P. Di Giorgio, P. du Castel, D. Dylus, P. Dzygiel, A. Felici, F. García-Alcalde,

A. Haldimann, M. Leipner, S. Leyn, S. Louvel, P. Misson, A. Osterman, K. Pahil, S. Rigo, A. Schäublin, S. Scharf, P. Schmitz, T. Stoll, A. Trauner, S. Zoffmann, D. Kahne, J. A. T. Young, M. A. Lobritz and K. A. Bradley, *Nature*, 2024, **625**, 566–571.
62. Q. Xiao and D. Pei, *J. Med. Chem.*, 2007, **50**, 3132–3137.
63. J. Nielsen, S. Brenner and K. D. Janda, *J. Am. Chem. Soc.*, 1993, **115**, 9812–9813.
64. M. C. Needels, D. G. Jones, E. H. Tate, G. L. Heinkel, L. M. Kochersperger, W. J. Dower, R. W. Barrett and M. A. Gallop, *Proc. Natl. Acad. Sci. U. S. A.*, 1993, **90**, 10700–10704.
65. M. D. Disney, L. P. Labuda, D. J. Paul, S. G. Poplawski, A. Pushechnikov, T. Tran, S. P. Velagapudi, M. Wu and J. L. Childs-Disney, *J. Am. Chem. Soc.*, 2008, **130**, 11185–11194.
66. S. P. Velagapudi, S. M. Gallo and M. D. Disney, *Nat. Chem. Biol.*, 2014, **10**, 291–297.
67. Q. M. R. Gibaut, Y. Akahori, J. A. Bush, A. Taghavi, T. Tanaka, H. Aikawa, L. S. Ryan, B. M. Paegel and M. D. Disney, *J. Am. Chem. Soc.*, 2022, **144**, 21972–21979.
68. D. F. Veber, S. R. Johnson, H. Y. Cheng, B. R. Smith, K. W. Ward and K. D. Kopple, *J. Med. Chem.*, 2002, **45**, 2615–2623.
69. M. Kansy, F. Senner and K. Gubernator, *J. Med. Chem.*, 1998, **41**, 1007–1010.
70. L. Di, E. H. Kerns, K. Fan, O. J. McConnell and G. T. Carter, *Eur. J. Med. Chem.*, 2003, **38**, 223–232.
71. P. Yu, B. Liu and T. Kodadek, *Nat. Biotechnol.*, 2005, **23**, 746–751.
72. J. Hu, A. I. Chan, E. Adaligil, I. Kekessie, M. Takahashi, A. Song, C. N. Cunningham and B. M. Paegel, *J. Med. Chem.*, 2023, **66**, 6288–6296.
73. P. Shieh, V. T. Dien, B. J. Beahm, J. M. Castellano, T. Wyss-Coray and C. R. Bertozzi, *J. Am. Chem. Soc.*, 2015, **137**, 7145–7151.
74. J. Morimoto, R. Amano, T. Ono and S. Sando, *Org. Biomol. Chem.*, 2019, **17**, 2887–2891.
75. B. Cai, A. El Daibani, Y. Bai, T. Che and C. J. Krusemark, *JACS Au*, 2023, **3**, 1076–1088.
76. L. K. Petersen, A. B. Christensen, J. Andersen, C. G. Folkesson, O. Kristensen, C. Andersen, A. Alzu, F. A. Sløk, P. Blakskjær, D. Madsen, C. Azevedo, I. Micco and N. J. V. Hansen, *J. Am. Chem. Soc.*, 2021, **143**, 2751–2756.
77. L. A. Schneider, B. Sauter, K. Dagher and D. Gillingham, *J. Am. Chem. Soc.*, 2023, **145**, 20874–20882.

CHAPTER 4

Assays of DNA-encoded Libraries Against Protein Targets on and Within Living Cells

SIAVASH SHAHBAZI NIA AND CASEY J. KRUSEMARK*

Department of Medicinal Chemistry and Molecular Pharmacology, College of Pharmacy, Purdue University and Purdue University Center for Cancer Research, 575 Stadium Mall Drive, West Lafayette, Indiana 47906, USA
*Email: cjk@purdue.edu

4.1 Introduction

The initial identification of hit compounds in drug discovery is a pivotal stage that includes many uncertainties and risks. The discovery process is complicated by the imperative to find hits that are not merely active, but also are synthetically tractable to enable improvements in selectivity, potency, and other drug-like characteristics. Identification of ideal drug hits poses a significant hurdle and emphasizes the need for robust strategies that enhance the likelihood of successful drug development.[1]

DNA-encoded chemical libraries (DELs) have emerged as an alternative approach to other screening methods and have much to offer in tackling drug discovery challenges. However, DELs are associated with many constraints. DELs offer much greater exploration of chemical space and lower assay costs compared to traditional screening methods.[2] By and large, molecules must be synthesized on unprotected DNA.[3] While initially this

Drug Discovery Series No. 85
DNA-encoded Library Technology for Drug Discovery
Edited by Guansai Liu, Casey J. Krusemark and Jin Li

Published by the Royal Society of Chemistry, www.rsc.org

was a significant drawback, more recent years have seen the development of greatly improved on-DNA chemistries to produce libraries with both high diversity and drug-likeness.[4] A major benefit of an encoded library is that an entire library can be assayed collectively in an affinity selection. This is typically conducted using a purified, tagged protein immobilized on a solid support to physically separate ligands away from non-ligands by their affinity for the target. Consequently, the conventional DEL approach can be constrained by the necessity of a purified protein with an affinity tag. To fully capitalize on the DEL benefits of highly complex libraries, assays are performed under conditions where the protein concentration drives the binding equilibrium to hit ligands in the library. Thus, there is a requirement for protein with high specific activity because it is the *active* protein concentration that sets assay sensitivity. In addition, all conformations of the protein (including inactive or non-native conformations) can potentially bind DEL ligands in an affinity selection, which can yield false positives. This polarity of the binding equilibrium is in direct contrast to conventional biochemical assays applied in HTS, where the small molecule ligand is present in excess. Thus, in HTS, it is the concentration of the small molecule that sets the assay sensitivity in the screen, and the presence of inactive protein is unlikely to affect assay results.

Extending the DEL approach to live cell assays would have many benefits over conventional, *in vitro* affinity-based selection assays with purified proteins. The main benefit is the ability to maintain the protein in its active, physiologically relevant state. Yet, a live cell approach has the challenges of protein targets being at modest to low concentrations within a complex mixture of many other proteins and biomolecules, which may contribute to assay background signal. Additional issues for DEL live cell assays arise due to the DNA tag. Recognition of DNA by cellular proteins may limit the accessibility of a DEL to the target protein. Also, activation of innate immunity pathways that sense exogenous DNA may alter the cellular state.[5] The DNA tag also hinders access to intracellular target proteins since it cannot readily traverse the plasma membrane. Despite these challenges, a few studies have reported successful discovery of ligands from DELs by targeting proteins on live cell surfaces or within live cells. These challenges underscore the need for innovative selection methods to effectively target proteins within the complex environment of live cells using DELs. In this chapter, we describe the different selection methods reported to date where live cells containing the target protein were used to assay a DEL. We discuss the challenges associated with each work, their potential advantages and disadvantages and, lastly, a future perspective. We have limited our discussion to in-solution, pooled DELs and omitted one-bead-one-compound DELs, which are covered in Chapter 3.

4.2 Live Cell-mediated DEL Affinity Selection for Cell Surface Proteins

Integral membrane proteins pose significant limitations for conventional DEL selection due to their associated challenges in the purification process.

Application of DELs to these proteins is highly desirable as G-protein coupled receptors (GPCRs) and ion channels are major drug classes.[6] A few successful reports have been reported targeting purified GPCRs, through protein stabilization by mutation,[7] detergents,[8] or nanodiscs.[9] While successful, the purification of an integral membrane protein is not trivial, and such approaches have the potential to alter the native protein structure.

4.2.1 Direct Equilibrium Binding to the Cell Exterior

In 2015, Israel and co-workers reported the first use of DELs with live cells and developed an approach to target the NK3 tachykinin GPCR.[10] Their selection design was a simple one and involved incubation of the DEL with cells and selection of ligands by separation of the cells by centrifugation (Figure 4.1A). Comparative analyses of multiple selections were key to this strategy. These included analyses of parallel selections against cells overexpressing the NK3 target protein, identical cells lacking expression of the target, and target expressing cells in the presence of a known NK3 antagonist, SB-235375 (K_i: 2.2 nM). Selection against NK3 receptor-negative cells and cells with a competitive free ligand would identify enriched molecules that may bind to other cell receptors or to cells non-specifically. In this study, the researchers employed HEK293 cells and expressed the target recombinantly using BacMam viral transfection. High NK3 receptor expression levels (~500 000 receptors per cell) were obtained in cells transiently transduced with the BacMam virus. Model selections were conducted and optimized using SB-235375. For test selections, a DNA-tagged SB-235375 was spiked into the DEL at varying concentrations spanning from 1×10^6- to 2.5×10^7-fold dilution. Across all dilutions, consistent enrichment of approximately 3000-fold was observed after three rounds of DEL selection, which validated that the binding was protein target-driven. As an additional control, the authors also prepared a fluorescently tagged test ligand on a DNA construct (SB-235375-DNA-Fluor) for use in a fluorescence binding assay. The assay revealed a 10–15-fold increase in fluorescence for NK3-expressing cells compared to the negative control, which was consistent with the enrichment measurements. The DEL selection assay was conducted using a three-cycle DEL comprising 41 million compounds centered around the 1,3,5-triazine core structure, as well as other non-triazine libraries. The selection yielded numerous, novel NK3 ligands, many with binding affinities in the low nanomolar range. Notably, compounds **1** and **2** (Table 4.1) emerged as standouts, showing K_i values of 0.6 and 1.2 nM, respectively.

Despite the great success in developing a live-cell method identifying novel ligands from DELs, certain challenges of this approach may limit its widespread adoption. The approach relies upon a high level of protein expression, which may be toxic to cells with other protein targets. By and large, the ligands discovered were of high affinity. Of course, this is desirable; however, if the library lacks ligands of high affinity, the selection may not yield hits. Ligands of low micromolar affinity are not likely to be enriched

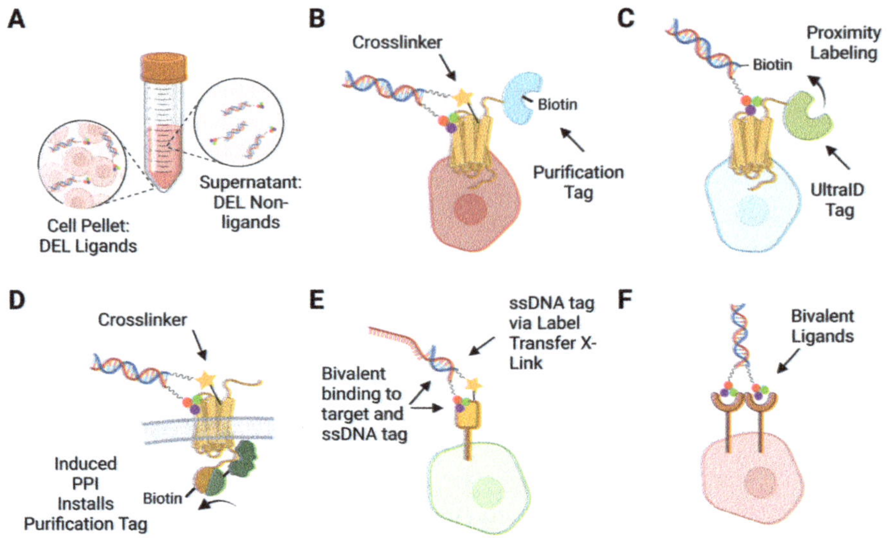

Figure 4.1 Approaches for selection of DELs against protein targets on the surface of live cells. (A) Selection by direct binding of the library to cells. Ligands are purified from non-ligands by centrifugation and washing of cells. Specificity to a target can be achieved by comparative analysis to selections to cells without recombinant expression of the protein of interest (POI).[10] Target agnostic selections can identify cell-selective binders.[41] (B) Crosslinking selection approach. Binding leads to selective affinity labeling and traps reversible binders covalently on the POI. After cell lysis, capture of the target *via* a fusion tag allows the purification of DEL ligands.[16] Alternatively, target agnostic selections can be applied with crosslinking and cell centrifugation as in (A).[11] (C) Selection by induced proximity to an enzymatic fusion protein on the protein target.[23] The UltraID tag on the POI enables acylation of amines in close proximity to biotin. Binding of the DEL ligand induces biotinylation of an appended amine on the DEL construct for subsequent enrichment with streptavidin beads. (D) Agonist-specific selection approach employing covalent crosslinking.[12] As with the approach in panel (B), DEL ligands are crosslinked to the POI by affinity labeling. Binding of agonists recruits downstream binding proteins, which leads to the reconstitution of a split protein tag, TurboID, that biotinylates nearby proteins. (E) Selection against endogenous protein targets using label transfer of an ssDNA tag.[13] A DNA-linked ligand (small molecule or antibody) to the POI is used to transfer an ssDNA tag to the POI in an affinity labeling step. The DNA-linked ligand is then removed by toehold displacement. Subsequently, the ssDNA serves to increase binding affinity to an ssDNA DEL by avid binding in a direct binding selection as in (A). (F) Bivalent DEL approach for targeting multimeric targets. The homodimerization of DELs allows improved enrichment in cell binding of DELs for an endogenous, dimeric protein target (CAIX) by avidity.[32]

Table 4.1 Representative active compounds identified by live cell DEL selections.

Compound	Target	Structure	Initial hit potency	Selection conditions	Library
1	NK3 tachykinin GPCR[10]	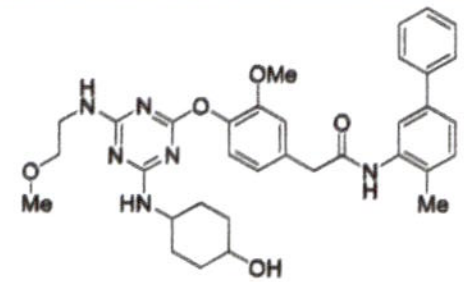	0.6 nM K_i (antagonist)	Centrifugation of cells overexpressing target	4.1×10^7 triazine
2	NK3 tachykinin GPCR[10]	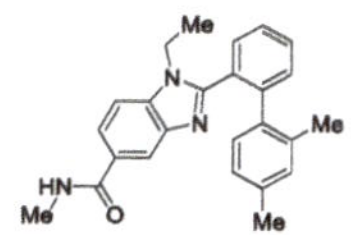	1.2 nM K_i (antagonist)	Centrifugation of cells overexpressing target	1.5×10^{10} pooled libraries
3	MDA-MB-231 cells (alpha-enolase)[11]	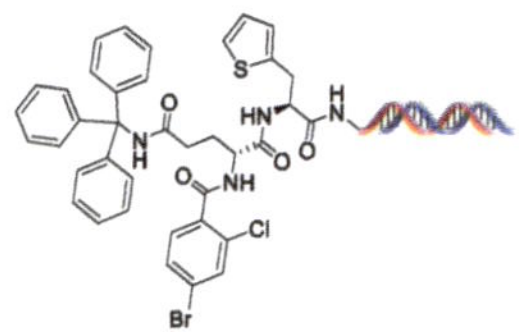	3.6 μM K_d	Crosslinking and centrifugation of cells	30.4×10^6 capped dipeptide
4	Kappa opioid receptor[12]	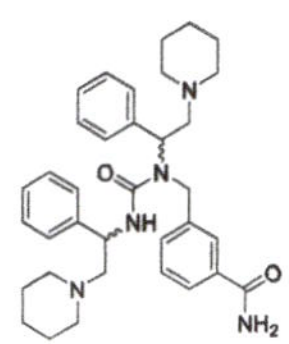	34 nM EC_{50} (biased agonist)	Crosslinking and split TurboID biotinylation	1×10^4 directed library

Table 4.1 (*Continued*)

Compound	Target	Structure	Initial hit potency	Selection conditions	Library
5	Folate receptor[13]	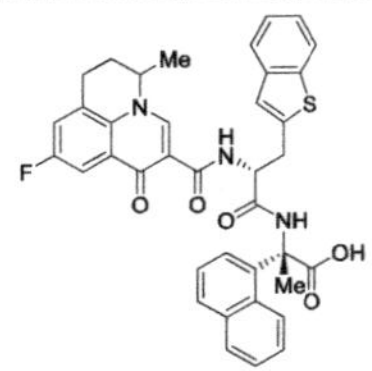	58 nM K_d	Centrifugation of cells with ssDNA-labeled target	30.4×10^6 capped dipeptide
6	Epidermal growth factor receptor[13]	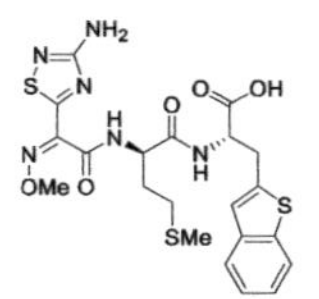	4.2 μM K_d	Centrifugation of cells with ssDNA-labeled target	30.4×10^6 capped dipeptide
7	p38α[14]	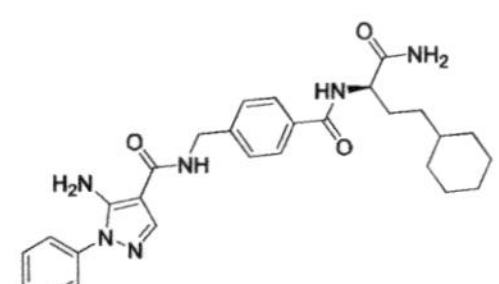	<25 nM IC_{50}	Injection into an oocyte and binder trap enrichment	194×10^6 pooled libraries

well in this approach; however, the authors did characterize some off-DNA hits with binding affinities as low as 900 nM.

In a recent work, Summerfield and coworkers expanded the application of cell-based DEL selection beyond merely identifying a target-specific hit molecule.[10] In this work, the authors applied the affinity selection of a DEL to live cells in a target-agnostic fashion. The authors used the centrifugation and washing of cells as a selection approach, as implemented by Israel *et al.*, with several murine and human cell lines. Their rationale was not to discover drug hits to specific targets but to develop a proteomic profile of the cell states as a method to potentially identify biomarkers and new therapeutic targets. As an initial test, a comparison of DEL binding to a mouse C2C12 myoblast cell line *versus* binding to differentiated myotubes was conducted. As expected, many non-specific binders were identified. Filtering based on enrichment features was able to identify several myotube-specific chemotypes, as well as dual myoblast/myotube binders. Following the identification of binders, their cell-specific binding capacity was confirmed by individually resynthesizing the compounds on-DNA and testing them against the C2C12 myoblast cell line and differentiated myotubes, as well as other cell lines, including BHK, A549, HEK293, and COS, to demonstrate specificity further.

Further experiments expanded DEL selections against several human cell lines with T and B cell lineages, including Molt4, Ramos, Raji, Daudi, and K562 cell lines. Compared to the initial myoblast *versus* myotube analysis, these immune cell lines have high morphological similarity, which suggests a greater challenge in identifying cell line-specific binders. The authors performed selection experiments against each cell line with multiple replicates and up to 6 independent experiments (different cell batches). Comparison of outputs from these experiments, particularly in the case of Molt4 and Ramos cells, suggested high reproducibility among replicates. Some experimental variabilities did arise among different cell batches that were impacted by the culture conditions (*e.g.* the interval of medium replacement, the number of cell passages, *etc.*). The selection results against these cell types yielded very low detectable signals and required a more advanced data treatment to identify cell-specific DEL ligands with statistical significance. They applied an in-depth analysis approach, DESeq2, that was designed for the determination of differentially expressed genes in biological samples.[15] The authors identified ~19 000 selective ligands to Molt4 cells from 8 distinct DELs and ~1000 selective ligands to K562 cells from 7 DELs. No cell-specific ligands were observed for Ramos, Raji, and Daudi cells. To confirm the cell-selective binding of the identified ligands, 118 and 94 predicted selective ligands were chosen for each of Molt4 and K562 cells, respectively, and were resynthesized on DNA for selection against all five cell lines. Out of the 118 Molt4 and 94 K562 ligands tested, 64 Molt4 and 42 K562 ligands were found to selectively bind to their targets with >10-fold enrichment compared to the background. Of note, only 1 Molt ligand and 6 K562 ligands were found to bind to one or more non-target cells. The authors also took advantage of on-DNA resynthesized ligands by assessing their binding in flow cytometry as an orthogonal assay using fluorescently-labeled DNA. Flow cytometry results supported the Molt4 specific binding results observed in the selection experiments. However, in the case of K562 ligands, none showed selective binding to the target cells, suggesting that the DEL selection enrichment and the flow cytometry assay were not translatable. Ultimately, based on the highest average signal from the on-DNA validation studies, a compound was synthesized off-DNA with a short PEG linker with a fluorophore at the location of the DNA tag. The compound was subjected to the binding assay with Molt4 cells at seven different concentrations using flow cytometry. Its K_d was found to be 100 nM and indicated Molt4 selective binding when compared with a negative control, an identified K562-selective binder. This result was further confirmed by a self-competition assay, where a 10-fold excess of the non-labeled Molt-4 selective ligand inhibited the binding of its fluorescently-labeled analog to 25% of the initial level, indicating target-specific binding.

4.2.2 Covalent Crosslinking to the Protein Target

Our lab reported the use of covalent crosslinking to address some of the challenges associated with DEL assays against live cell surface targets.[16] Such approaches have improved selection outcomes for discovery of nucleic acid aptamers with both formaldehyde and photocrosslinking.[17] In a live cell scenario with a DEL, the combination of modest ligand affinity and low target protein concentration results in a low proportion of DEL ligands bound to the target protein at equilibrium. Covalently trapping the equilibrium through covalent crosslinking can improve selections.[18,19] Selective purification of ligands from the DEL to the target protein of interest can be conducted subsequently by a recombinant tag on the protein.

In this work, we used a previously demonstrated approach to affinity labeling with DNA-linked ligands where a covalent crosslinking group is presented toward the target protein on the 3′ end of the DNA strand opposite a DEL member on the 5′ end (Figure 4.1B).[20] To apply this assay, we tested this approach using a model GPCR protein target, the delta opioid receptor (DOR), expressed recombinantly on live HEK293T cells. The receptor was expressed as a fusion protein with the SNAPtag, which allowed purification on streptavidin beads after covalent labeling with a benzylguanine-biotin.[21] To optimize selection conditions, we used a known high-affinity tri-peptide ligand, Dmt-Tic-Lys (DTK, 2,6-dimethyl-L-tyrosine (Dmt), 1,2,3,4-tetrahydroisoquinolone-3-carboxylic acid (Tic)) (off-DNA K_i: 150 pM). Initially, this ligand was prepared on fluorescently-tagged DNA, which was then incubated with HEK-293T cells expressing SNAPtag-DOR. Imaging of these cells demonstrated the retention of the fluorescent signal, indicating the necessity of both the ligand and the crosslinker (Figure 4.2). The requirement for the crosslinker in these tests suggested the affinity of the

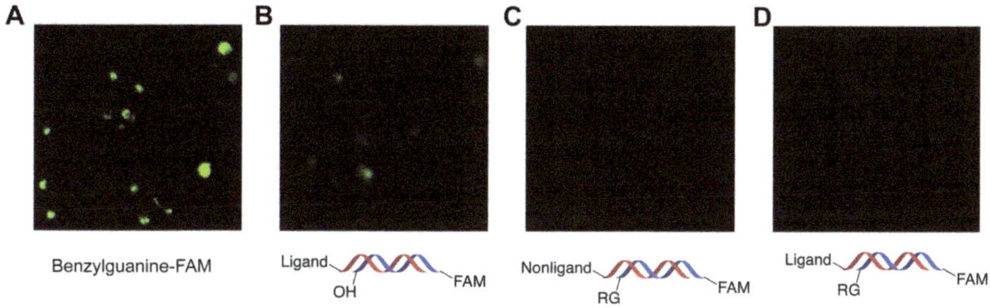

Figure 4.2 Labeling of a cell surface protein (δ opioid receptor (DOR)) on HEK-293T cells with a DNA-linked ligand. Cells were treated with 50 nM DNA construct or benzylguanine-FAM for 2 hours and then washed 3 times with buffer prior to imaging. FAM = fluoresceine amide; ligand = Dmt-Tic-Lys (DTK); RG = reactive group (sulfonyl fluoride); nonligand = unmodified DNA. Adapted from ref. 16 with permission from American Chemical Society, Copyright 2019.

DNA-linked DTK ligand was likely reduced significantly by tagging with DNA. In test qPCR selection assays using a suspension HEK cell line to increase cell density, a 180-fold higher enrichment was observed for the DTK ligand compared to the DOR non-binder control. The efficacy of the DNA recovery was poor with only 0.05% of the input ligand DNA being recovered. Then, two different peptide ligands, Dmt-Pro-Phe-Phe (K_i: 28 nM) and Tyr-Tic (K_i: 240 nM), were used as parental structures to design a 96-membered, scanning positional library peptide DEL, which was assayed against DOR expressing cells. We synthesized six unique hit compounds off-DNA and assessed their binding affinities through a radiolabeled ligand displacement assay. The K_i values obtained from these assays were in line with the results observed with DEL selections. The low recovery of the process would likely preclude the application of the assay with high complexity DELs. Poor recovery was attributed to difficulty in the solubilization and quantitative streptavidin capture of the receptor labeled with the DNA construct. Microscopy experiments comparing fluorescent labeling of the receptor with FAM-labeled DNA conjugates to benzylguanine-FAM labeling of the SNAPtag suggested the affinity labeling step was reasonably efficient (Figure 4.2).

In a recent preprint study, Li and coworkers also implemented a selection approach based on photocrosslinking-mediated stabilization of the ligand–receptor interactions on the cell surface.[11] They performed DEL selections with live cancerous cells without predefined targets. They aimed to identify novel cell surface-targeting ligands for use in cancer diagnosis and targeted therapeutics. Initially, the authors optimized their selection conditions using aryl sulfonamide ligands and carbonic anhydrase 12 (CA-12), an endogenous target highly expressed on A549 cells. Their approach involved hybridization of an aryl azide-containing 15-mer oligonucleotide to an ssDNA DEL. Photoaffinity labeling to cell surface receptors would trap binders on the cells, which are selectively isolated by centrifugation and washing of the cells. Heat treatment of the cells was used to elute the binders from the cells by denaturation of the DEL from the crosslinked strand. Test selections with the modest affinity ligand, carboxybenzyl sulfonamide (CBS (~1 μM K_d)), gave 200-fold enrichment after two rounds of selection. Doping the CBS-DNA into a 67 600-membered DEL gave a clear 40-fold enrichment that was dependent on the use of photocrosslinking.

To target their cancer cell lines, the authors utilized a 105-million-membered capped tripeptide DEL that was traditionally prepared using "headpiece" DNA. The library was converted to single-stranded DNA by lambda exonuclease digestion to allow hybridization to the aryl azide crosslinking DNA.[22] Selections were performed with the DEL against both MDA-MB-231 (highly metastatic) and MCF7 (lowly metastatic) breast cancer cell lines. Considering the possibility of binding of multiple ligands to multiple targets on the cell surface leading to enrichment, a disynthon-based approach was used for data analysis. Several disynthon clusters were enriched in the MDA-MB-231 cell selection specifically relative to the MCF7 cell selection. Identified cluster 1 was highly enriched and contained 420 compounds. This cluster was resynthesized on non-encoding 16-nt ssDNA and hybridized to a fluorescein-containing complementary oligo. As quantified by flow cytometry, enhanced fluorescence was observed associated with MDA-MB-231 cells, while the MCF7 cells and control DNA did not show a noticeable fluorescence. Titration of the on-DNA ligands allowed an estimate of the collective binding affinity of the cluster to the cells (~17 μM (K_d)). Further optimization gave a cluster collection of 168 compounds with higher affinity (K_d: 4.1 μM). Using this compound set, the authors demonstrated cell-selective delivery of a drug, doxorubicin, and a photosensitizer that enabled cell death through photogenerated reactive oxygen species. Seven hit compounds from this set were individually synthesized on DNA, which showed affinities for MDA-MB-231 cells ranging from 3.0 to 42 μM K_d. One hit compound (compound **3**, Table 4.1) was used for photoaffinity labeling and purification of the receptor. After proteomic analysis, α-enolase (ENO1) was identified as a likely cell surface receptor.

Both studies[11,16] demonstrated the ability of covalent crosslinking to enable enrichment of modest to low affinity ligands to targets on the surface of live cells. The drawbacks include the necessary appending of a crosslinking group to the DEL. While this can be included within the initial DEL build, it is not trivial to append a suitably positioned crosslinker to existing DELs. The approach used by Li *et al.* to digest one strand of a headpiece library followed by addition of a crosslinker by hybridization of a modified oligo should be generally applicable to most DELs.[22] While the efficiency of crosslinking is likely target dependent, these studies suggested crosslinking yields in the DEL context can be sufficient. For our work, we achieved target specificity through recombinant expression of the target with a purification tag (SNAPtag); however, the purification step proved challenging and led to low recovery of ligand DNAs. Achieving specificity through comparative analyses to selections with non-expressing cells or cells in the presence of an excess of a known ligand may help to address the recovery issue.

4.2.3 Proximity-labeling with an Enzymic Fusion Protein Tag

Our lab explored an alternative proximity-labeling approach for live cell DEL assays and demonstrated its use with live cells and the DOR model receptor system.[23] This approach involved the expression of the target protein as a fusion protein to the engineered biotin ligase, UltraID, a modified version of the commonly used BioID tag for proximity labeling of interacting proteins (Figure 4.1C).[24] Using biotin and ATP as substrates, the UltraID tag generates reactive acyl phosphate biotin groups (biotinoyl-adenylate) that can react with amines (*e.g.*, lysine side chains) within a ~10 nm radius. For the live cell DEL assay, the DOR was expressed as a fusion protein to UltraID on the cell exterior, and a free amine was added to the DEL construct on the DNA strand opposite of the ligand. Binding to the target protein leads to selective, proximity-induced biotinylation of the amine on the DNA, which is subsequently purified from non-ligands with streptavidin beads (Figure 4.1C).

The benefits of this approach include removing the need for purification of the protein and signal amplification through enzymatic turnover of the tag. Test selection assays with the DOR and a high-affinity peptide ligand yielded improved recovery of ligand DNA (~1%) with similar fold enrichments (300-fold) to crosslinking-based selections. The drawbacks of the UltraID tag include the potential for covalent modification of the target protein with biotin prior to the assay, which could lead to non-native structure and selection artifacts. A proximity-based decaging of a biotin label using a nanoluciferase tag was also explored in this work, which would avoid this issue. Live cell assays using constructs containing photocaged biotin were unsuccessful, presumably due to the hydrolytic instability of the photocaging group. Improved photocages and/or alternative enzymatic tags for proximity-based labeling are likely to improve the utility of this strategy.

4.2.4 Covalent Crosslinking with Split Protein Reconstitution for an Agonist-specific Selection

Our lab additionally combined a proximity-labeling approach with affinity-based crosslinking to enable a cell-based DEL selection that was specific for GPCR agonists.[12] In this work, we utilized split protein reconstitution of a fusion protein to enable labeling of the receptor with a purification tag (biotin) only if the bound DEL ligand agonized the receptor and recruited downstream protein binding partners (Figure 4.1D). This work utilized the opioid receptor family as a model system, yet the approach should be broadly useful for direct enrichment of ligands that induce protein–protein interactions. This work is built upon prior work with GPCR assays that have used split protein tag reconstitution to assay the recruitment of specific signaling proteins. Such assays have included split versions of GFP,[25] beta-galactosidase,[26] and nanoluciferase.[27] Our approach used a split TurboID tag, another variant of the engineered biotin ligase (BioID) proximity labeling tag, which had demonstrated utility with other GPCR systems.[28] As with the UltraID tag, TurboID leads to the biotinylation of nearby proteins through the acylation of lysine amines. Combining agonist-induced labeling with affinity crosslinking provided a way for selective purification of DEL ligands that agonize the receptor. We initially validated the assay with a model peptide agonist for the DOR, Met-Enkephalin-RF, which shows arrestin bias in signaling.[29] Using HEK cells transiently transfected with the DOR linked to the C-terminal portion of the TurboID and arrestin linked to the N-terminal portion of TurboID, a significant increase in GPCR biotinylation with agonist treatment *versus* an antagonist was demonstrated by both streptavidin blot and microscopy. Test selection assays using on-DNA model ligands yielded a 30-fold enrichment of the Met-Enk-RF agonist and no enrichment of the potent DTK antagonist. Again, the efficiency of the process was poor with the recovery of the agonist being approximately 1 part in 10 000 (0.01%). We additionally tested model selection assays for G-protein signaling using the mu-opioid receptor (MOR). This involved the recruitment of a camelid nanobody, nanobody 39 (Nb39), which is specific for the G-protein activated state of opioid receptors, as a proxy for activation of G-protein signaling.[30] In this assay, the agonist peptide yielded 80-fold enrichment and 0.02% recovery. Subsequently, the authors prepared a targeted DEL of 9216 members and applied the assays to select for both Nb39 and arrestin recruitment against 3 opioid receptor targets. Particular selections gave unique enrichment profiles. Fold enrichments were modest, ranging from maximally 3-fold to 20-fold enrichment, depending on the assay. Based on their unique enrichment profiles, 6 compounds were selected for off-DNA synthesis and testing. Only one of these was active in cell-based assays, but all showed at least modest activity in gel-based biotinylation assays, suggesting low potency. To increase the probability of discovering potent agonists, the authors performed binding selections using the recently developed enzyme-mediated proximity approach with the UltraID tag.[23] Using enrichment data for both the binding assay and recruitment assays, four additional compounds were selected for off-DNA synthesis, which all showed agonist activity in cell-based assays. The selectivity and signaling bias of off-DNA compounds were not perfectly reflected in the selection data, which may be expected given the noise level of the data. Two reasonably potent, G-protein biased agonists, [66, 66] ($EC_{50} = 100$ nM) and [71, 66] ($EC_{50} = 34$ nM) (Table 4.1, compound **4**), for the kappa opioid receptor (KOR) were discovered, which showed selectivity and signaling bias profiles that were consistent with DEL data.

The main benefit of this approach is the unique ability of a selection to directly assay a specific biochemical function, in this case GPCR receptor agonism. The method would be generally applicable to other receptors that act through the recruitment of intracellular proteins. The low efficiency and fold enrichments were a drawback and were largely attributed to the same difficulties observed in the poorly efficient purification of a DNA-labeled membrane protein in our previous work.[17] These difficulties were compounded by the complexity of the split protein tag reconstitution. Previous studies with split TurboID with GPCRs and arrestin recruitment demonstrated high "switchability" and no detectable biotinylation of the receptor in the absence of an agonist.[28] In our work, however, we required an extended protein expression period (~48 hours) to allow protein concentrations to accumulate to high levels. During this time, we observed background biotinylation of the receptor, which was likely due to the low-level affinity of the two TurboID halves. This limited the signal to noise in distinguishing agonist from antagonist ligands. Replacing the split protein reconstitution approach with a downstream binding partner labeled with a turn-on, proximity labeling fusion protein, such as APEX,[31] could improve this method.

4.2.5 Endogenous Protein Targets *via* Label Transfer of an ssDNA Tag

Another approach for DEL assays of cell surface receptors was developed by Xiaoyu Li and coworkers and also exploits affinity labeling.[18] This approach allows targeting of endogenous proteins and does not require the recombinant expression of a purification tag on the target protein. This approach does require a known ligand or antibody to the target protein, however, which is used to achieve target specificity. The method involves an initial photoaffinity labeling of the target protein using a known, DNA-linked ligand, to place a DNA tag on the target (Figure 4.1E). Importantly, this labeling step is conducted using an excess of DNA-linked ligand/antibody to saturate the binding sites on the cell surface. In this first step, affinity labeling is conducted in the dual display format with the ligand on one DNA strand and the photocrosslinker (aryl azide) on the opposite strand. By using a longer DNA strand on the known ligand, the ligand strand can be removed from the protein after labeling by using DNA hybridization in a toehold displacement step. This leaves the target protein uniquely labeled (among other cell surface proteins) with an ssDNA tag. This ssDNA tag serves to increase the overall binding affinity of ligands within an ssDNA DEL by avid binding to the ligand binding site on protein and the DNA tag through Watson–Crick base pairing. This "affinity boost" is easily tunable by modulating the number of hybridizing base pairs between the tag and the DEL. The assay of a DEL is subsequently conducted simply by binding of an ssDNA DEL to labeled cells and removal of non-ligands by centrifugation and washing.

In this work,[13] the authors optimized the assay using the folic acid (FA)/folate receptor (FR) system with HeLa cells, which endogenously express FR. Using a DNA-linked FA (~700 nM K_d to FR) and aryl azide photocrosslinking, they demonstrated selective and ligand-dependent label transfer of an oligonucleotide tag to FR using microscopy, cell sorting, immunoblotting, and mass spectrometry. Selection of a 4800-membered DEL containing the FA control construct was conducted using seven complementary bases between the DEL and the ssDNA label. DNA sequencing revealed selective enrichment of the FA construct of approximately 50-fold. Selections using cells without the ssDNA affinity label did not show enrichment. Similar control selections were performed with aryl sulfonamide ligands to carbonic anhydrase 12 (CA-12) on the surface of A549 cells. These tests yielded more robust enrichments (2000- to 3000-fold), perhaps due to the higher expression levels of the receptor (~30 nM). Subsequently, a 30 million-membered capped dipeptide DEL was similarly applied with the assay against FR, and novel FR ligands were discovered. Five hits with enrichment values all around 50-fold were selected for synthesis and characterization off-DNA. All were confirmed as ligands with an affinity range of 60 nM (compound **5**, Table 4.1) to 25 μM. Impressively, these selections were conducted with cells that did not overexpress the target receptor. The concentration of the protein target in the selection was estimated using flow cytometry to be ~5 nM. The authors additionally calculated the theoretical affinity increase through DNA hybridization to seven base pairs to be ~19 000-fold, which was critical to enriching such modest affinity ligands under low protein concentrations.

In the same paper,[13] the authors extended this approach to demonstrate label transfer of the ssDNA tag by using DNA-linked antibodies to target proteins rather than small molecule ligands. The authors randomly derivatized antibodies using a heterobifunctional crosslinker containing both *N*-hydroxysuccinimide and maleimide groups followed by linkage to thiol-containing oligonucleotides. The generality of the antibody-based label transfer of oligonucleotides was demonstrated with four receptors: epidermal growth factor receptor (EGFR), FR, transferrin receptor 1, and CD44. They found that extending the length of the oligo label to 18 bases improved labeling efficiency and minimized self-labeling of the antibody. In the case of EGFR, the labeling efficiency was assessed by a gel shift-immunoblot analysis and was found to be quite efficient with about 25% of the total protein being oligo-labeled. They went on to apply the affinity-boosted selection of their 30-million-membered DEL with EGFR. The top five enriching molecules (80–110-fold) were characterized off-DNA, and all were confirmed as ligands with affinities ranging from 4 μM (Table 4.1, compound **6**) to 70 μM.

There are significant benefits of this approach relative to other cell surface selection methods. Uniquely, the use of the ssDNA tag for avid binding allows modulation of equilibrium binding affinity, which should enable targeting of proteins at very low concentrations. This approach also allows targeting of endogenous, protein targets with target specificity. The approach does, however, require an existing ligand or antibody to the target. The specificity for a particular target in the DEL selection will then reflect the specificity of the ligand or antibody. Also, performing the label transfer step does add complexity to this approach, particularly in the case of the antibody-mediated labeling. Antibodies with unknown binding sites may not place an ssDNA label suitably positioned for avid binding with DEL ligands at the desired binding site.

4.2.6 Bivalent DEL Binding to a Multivalent Protein Target

Neri and co-workers conducted a careful study of a cell-based selection to optimize selection conditions and investigated the effects of bivalency (Figure 4.1F).[32] In this work, they performed a direct cell binding selection through centrifugation and washing to target carbonic anhydrase IX (CAIX), which is a homodimeric cell surface protein often overexpressed in tumor cells. To optimize selection conditions, the authors prepared both monovalent and bivalent control ligand constructs with the high-affinity CAIX ligand acetazolamide (AAZ) ($K_d = 8.7$ nM). The enrichment and recovery of these constructs relative to a negative control were evaluated by qPCR in selections with varying blocking reagents, salt concentrations, total cell number, number of cell washes, and association time. Conditions with SKRC-52 tumor cells indicated a 5-fold enrichment and 2.7% recovery for the monovalent AAZ-DNA, while a significantly higher enrichment of 65-fold and a recovery of 34% were observed in the case of bivalent AAZ-DNA. Conditions where the washing buffer contained an increased salt concentration (500 mM NaCl) yielded ~5-fold increases in enrichment for both monovalent and bivalent constructs, presumably through the reduction of non-specific ionic interactions with the DNA and the cell surface. It was also found that the number of cells used in selection had minimal impact on the results above a minimum threshold of 3 million cells. Additionally, inclusion of different blocking agents (herring sperm DNA, bovine serum albumin (BSA), or milk) yielded modest or no influence on selection performance. Overall, these findings suggested that the bivalent ligand display consistently exhibited markedly enhanced selection performance. To further confirm the results obtained by using SKRC-52 cells, the authors also performed selections with CT26.3E10 cells expressing CAIX and on CT26.wt (with no CAIX expression) where similar results were achieved. Enrichment was dependent on the presence of the CAIX target protein.

With the optimal experimental conditions in hand, the authors utilized a DNA-encoded self-assembled library with one million members to perform a selection on cells and compared the results with a traditional affinity selection with CAIX on beads. The library contained AAZ and additional aromatic sulfonamides with CAIX K_d values in the nanomolar to sub-micromolar range. The experiments were performed on beads with/without CAIX, CT26.3E10 cells expressing CAIX and CT26.wt (negative control), and CAIX-positive SKRC-52 cells. Among the enriched library members, there were four bivalent binders identified as the most abundant members with enrichment factors up to 11 500, which were also enriched in selection experiments performed on CAIX expressing cells, but not the CT26.wt cells (CAIX negative) or for beads without CAIX. Despite this, there were clear differences between the selections on beads *vs.* the cells. The cell selections demonstrated a higher background noise and lower enrichments for both monovalent and bivalent binders. Additionally, in the selections involving CAIX-expressing CT26.3E10 cells, only monovalent AAZ derivatives and the bivalent sulfonamides were detectable. Selections with the CAIX-positive SKRC-52 cells additionally showed enrichment of a 4-chloro-3-sulfamoylbenzoic acid (Cl-SABA) derivative, a modest affinity CAIX binder ($K_d = 0.5$ μM). Of note, in the case of selections with CAIX-positive SKRC-52 cells, a higher relative abundance of bivalent and monovalent CAIX binders, as well as lower background noise, was achieved with a higher library input (5× instead of 1×10^6 copies per member per selection) and shortening the library incubation time (1 minute instead of 1 hour). The benefits and drawbacks of this approach are largely shared with previous work with binding assays by centrifugation and washing of cells. By capitalizing on avidity, the authors increase the binding of low-affinity monomers in their bivalent DELs to this dimeric protein target. While many cell surface proteins are multimeric, this approach would not facilitate selection to monomeric targets.

4.3 Live Cell-mediated DEL Affinity Selection for Targets Within Cells

4.3.1 Cell Penetration by a Cell Penetrating Peptide Followed by Covalent Crosslinking to the Protein Target

Our group extended the selection approach to allow DELs to target proteins within live cells.[16] Again, this approach relies on the induction of a covalent linkage between the DNA and the protein targets *via* affinity labeling. Delivery of the DEL into the cell enables the affinity labeling step to assay the binding of a DEL member to a target protein within the live cell. Subsequently, cells are lysed, target proteins are captured *via* a recombinant fusion protein tag, denaturing washes can be applied, and the associated enriched ligands are identified by DNA sequencing.

In this study, delivery of the DELs into live cells was marked as a challenging process. To overcome this, we appended a cyclic, cell penetrating peptide (cCPP12) (Figure 4.4A),[33] which has demonstrated high cell entry and endosomal escape efficiency, to each DEL construct (Figure 4.3A). Cellular uptake of fluorescently labeled mock DEL constructs was confirmed by microscopy in human mammary epithelial cells (HMLEs) (Figure 4.4A). qPCR analysis additionally validated cell entry and was also used to determine the effect of DNA length on cell penetration, which showed lower cell entry with longer DNA constructs. To validate the delivery of the modified DNAs into the cell cytosol (and not trapped within endosomes), we used the HaloTag-based chloroalkane penetration assay (CAPA).[34] Also, target engagement of two control DNA-linked ligands to two recombinant cytosolic targets was shown by gel analysis using ligand–DNA constructs additionally containing a fluorescein tag (Figure 4.4B). Incubation of the suspension adapted HEK293 (Freestyle 293-F) cells, followed by washing and SDS-PAGE buffer-mediated lysis, together with the addition of excess free ligand, confirmed

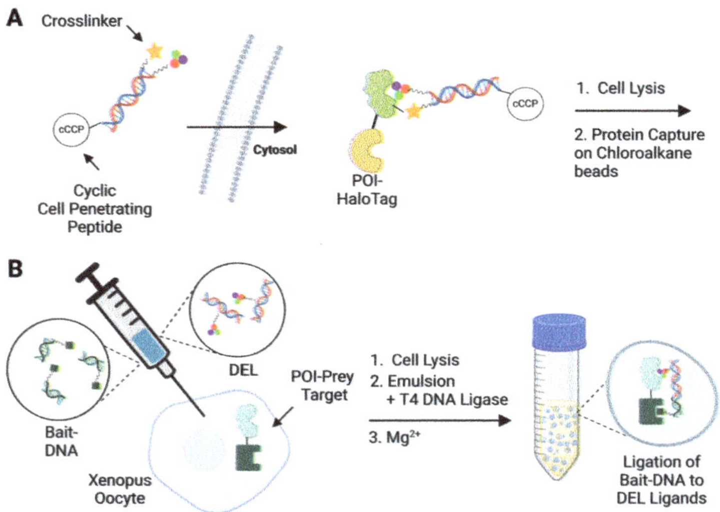

Figure 4.3 Approaches for selection of DELs against protein targets within live cells. (A) Selection by appending a cyclic cell penetrating peptide and crosslinking to a HaloTag-POI.[16] A cell penetrating peptide allows delivery of a DEL construct into the cellular cytosol. Covalent crosslinking of DEL ligands to the POI enables capture after cell lysis by a HaloTag fusion protein. (B) Selection by injection of a DEL into a *Xenopus* oocyte.[14] Injection of a DEL along with bait-DNA allows binding of both DEL ligands and bait-DNA to the POI-prey target recombinantly expressed in the oocyte. After cell lysis, emulsion formation enables co-compartmentalization of DEL binders and bait-DNA for a selective ligation of the bait-DNA to the DEL. Subsequent PCR amplifies only ligated products by using the bait DNA sequence as a primer.

that the labeling occurred within the cells. Labeling was shown to be dependent on the presence of cCPP12, the ligand, and the sulfonyl fluoride crosslinker on DNA.

Upon successful labeling experiments, a test selection assay was performed against HaloTag-chromobox protein homologue 7 chromodomain (CBX7-ChD) with a known DNA-linked peptidic ligand, 4-BrBA (K_d: 270 nM off-DNA). The HaloTag fusion protein allowed covalent capture of the protein after cell lysis using chloroalkane beads. HEK293 cells adapted to suspension culture (FreeStyle 293-F) were found to be best suited for selections. These cells lack the cGAS–STING pathway that detects cytosolic DNA and are readily handled at high cell density. The selection experiment involved the use of controls, including a no ligand construct and a ligand construct that lacked the cell penetrating peptide. To ensure that the binding equilibrium was protein-driven, the ligand constructs were added to cells at a concentration (1 nM) well below the K_d to the target. By qPCR analysis, we reported a 20-fold enrichment relative to the non-ligand control, which was dependent on the presence of the cCPP. An enrichment of 1200-fold was observed in analogous experiments in cell lysates, suggesting that cCPP-mediated delivery is an inefficient method to achieve DNA access to the cytosol. Similarly, the cell-based selections gave poor recovery of the ligand DNA with only 0.01% of the input being recovered. In contrast, cell lysate ligand recoveries were 4%. This marked difference is likely due to a

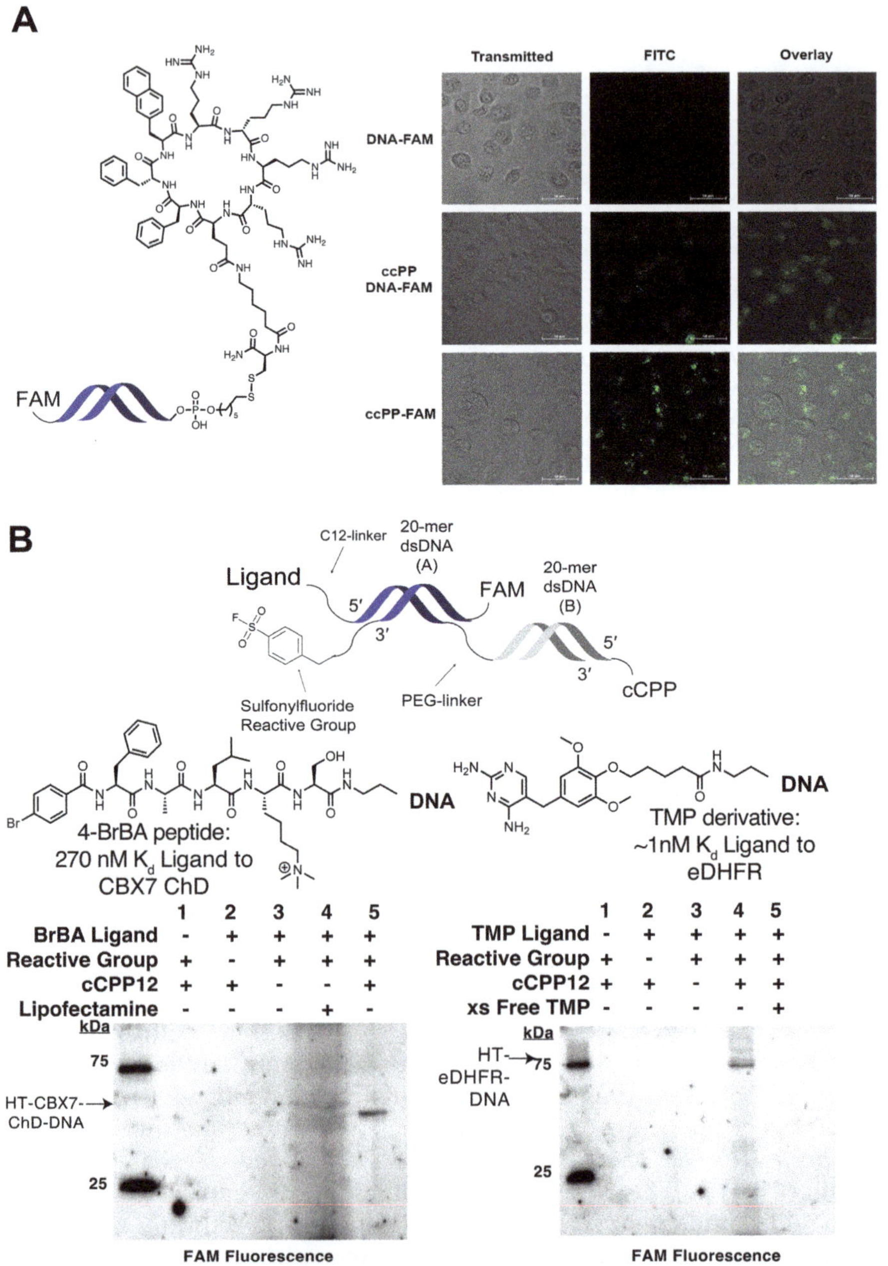

combination of factors, including poor cell entry (estimated at ~10% for similar length DNA), poor endosomal escape, and DNA degradation in the live cells.

With demonstrated selection conditions, we performed a selection of a 96-membered DEL against CBX7-ChD, which contained several known ligands and expected SAR. Cell selection results were consistent with previously established binding SAR. Control ligand constructs that were doped into the library selection reproduced qPCR results and showed enrichment that was dependent on the presence of the cCPP. A comparative selection was conducted with both pure protein and with the HEK cell lysate, which showed strong correlations to the cell-based selection. Ten hit molecules with a range of enrichment values from the library were selected for off-DNA synthesis and validation. IC_{50} values in a ligand displacement assay showed consistent affinities to observed enrichments from sequencing.

A major benefit and rationale of this approach is the ability to target proteins within live human cells, which is a more relevant environment. While low fold enrichments and recoveries would likely limit this approach to small libraries, the extreme sensitivity of DNA detection enables detection of binding and would allow utility with smaller targeted libraries. The ability to engage targets in live cells with DNA-linked small molecules was a critical demonstration and should enable discovery with challenging targets with further development. Improvements in cell entry, in particular, would facilitate this approach. While cytosolic access of DELs was demonstrated, targets within other cellular compartments are not likely accessible due to the DNA tag.

4.3.2 Direct Injection Into a *Xenopus* Oocyte Followed by Binder Trap Enrichment

Hansen and colleagues introduced another DEL selection technique conducted directly within live cells.[14] This approach circumvents the necessity of employing cell-penetrating peptides for delivering DELs into cells. Here, DELs are delivered by direct injection into the cells of *Xenopus laevis* oocytes

Figure 4.4 Delivery of DNA-linked ligands into cells. (A) Structure of the cCPP and linkage to DNA for constructs used in microscopy. Cell penetration of DNA constructs by the cyclic cell penetrating peptide (cCPP). 20-base pair DNAs containing either a FAM label only or both a cCPP and a FAM were incubated with HMLEs at 2 μM for 4.5 hours prior to imaging. A direct conjugate of the cCPP to FAM without DNA was used as a control. (B) Target engagement of DNA-linked ligands in Freestyle 293-F cells expressing POI-HaloTag fusions. Sulfonyl fluoride-mediated affinity labeling shows binding of DNA-linked ligands to CBX7 and eDHFR in live cells. Fluorescence imaging of SDS-PAGE of cell lysates showing selective labeling of targets at expected molecular weights. Adapted from ref. 16 with permission from American Chemical Society, Copyright 2019.

with a microinjection apparatus. Notably, these oocytes boast a substantial size (approximately 1 mm in diameter), making them over 100 000 times larger than typical mammalian somatic cells. Moreover, they are amenable to heterologous protein expression, rendering them an ideal platform for this purpose.[35] To discern binding to the recombinantly expressed target protein, the authors utilized the cellular binder trap enrichment (cBTE) technique (Figure 4.3B).[36] In this method, the protein of interest (POI) is tagged with a small protein, referred to as the "prey," possessing a high affinity for a small molecule known as the "bait" that is linked to a DNA tag. For the prey protein, they used the catalytic domain of carbonic anhydrase IX and a high-affinity aryl sulfonamide ligand. For this assay, a DEL and bait-DNA are injected into oocytes that are expressing the POI-prey target protein. After incubation, the cells are lysed, and the lysate is subjected to a water-in-oil emulsion formation with a surfactant and a solution containing the T4 DNA ligase. If a DNA molecule binds to the target, both the library member and the target protein with its DNA tag will be entrapped within the same droplet. Non-ligand molecules are likely to be within droplets lacking a POI-prey-DNA tag. Treatment of the emulsion with Mg^{2+} activates the ligase, leading to a selective ligation between the tag DNA and the DEL binders. Since the protein-linked DNA tag contains the PCR priming site, DEL binders are selectively amplified in a PCR after disruption of the emulsion.

Using this approach, the authors assayed a 194-million-member DEL against three therapeutic targets: mitogen-activated protein kinase 14 (p38α), human acetyl-coenzyme A synthetase 2 (ACSS2), and human dedicator of cytokinesis 5 (DOCK5, Dock Homology Region 2). To prevent degradation of the bait-DNA as well as the DEL in the oocyte, phosphorothioates were introduced at the ends of the DNA constructs. The assay was performed by running nine different experiments in parallel: the three POIs, each with three different bait-DNA concentrations. The results identified 323 structurally distinct hit molecules. Of the 323 hits, 154 were specific for p38α, 20 for ACSS2, and 76 for DOCK5. The observed enrichment values of top hits were in the range of 100–1000-fold. Selections were conducted with a copy number of just 6000 molecules, each suggesting that the selection method is highly efficient. For a given protein target, 30–50% of the hits were observed in all three bait-DNA concentrations suggesting robustness and good reproducibility. The authors selected 21 p38α hit compounds with different physicochemical properties and enrichment signals for validation in an enzymatic assay of ADP accumulation. Of the 21 compounds tested, 15 showed IC_{50} values below 10 μM and 9 below 1 μM (*e.g.*, compound 7, Table 4.1).

Overall, Hansen and colleagues reported a live cell DEL selection method that demonstrated excellent success and robustness in identifying the hit compounds from a complex library. The method is not without its limitations. A drawback may be the need for specialized equipment and expertise for cellular injection. Also, the low copy number used (6000 per molecule) and the volume restrictions (50 nL injected into a 1 μL cell volume) may suggest limitations for extending this to ultracomplex DELs ($\sim 10^{12}$ diversity), as it may not be feasible to inject a large number of cells. For some human protein targets, it may not be possible to replicate physiologically relevant conditions within the *Xenopus* system. Also, given the time scale of cell lysis and emulsion formation of the binder trap step, most ligands of modest affinity will dissociate from the protein target and rebind within that time frame, which would indicate that the method does not assess binding within a live cell but in the lysate. Combining cell injection with either crosslinking or proximity labeling, rather than the binder trap, would be straightforward and would address this issue. For many targets, these concerns are inconsequential, and the track record of *Xenopus* oocytes as a robust platform for human protein expression supports general applicability. Due to the use of these cells for protein expression, *Xenopus* oocytes are readily available commercially.

4.4 Perspective and Outlook

Despite the challenges associated with DEL selections with targets within and on live cells, several new bioactive compounds have been discovered in this fashion (Table 4.1). The levels of robustness for the presented assays vary. Successful demonstrations have been provided from industry groups with highly complex libraries from both Israel *et al.*[10] (10^{10} complexity) and Hansen *et al.*[14] (10^8 complexity). In fact, live cell DEL assays within *Xenopus* oocytes are offered commercially.[37] Other demonstrations range in their potential utility with highly complex libraries. The utility of these assays in the broader area of drug development extends beyond initial hit discovery. These aspects are just beginning to be explored. Agnostic selections against live cell surfaces offer possibilities for proteomic profiling and the discovery of new drug target proteins, biomarkers, and small molecule ligands for drug targeting and imaging.[11] Also, targeted libraries of lower complexity offer great potential for on-DNA medicinal chemistry.[38] This may include cellular target engagement, optimization of ligands for affinity and selectivity, distinguishing receptor agonists and antagonists, and measuring the signaling bias of agonists.

Particularly for cell surface targets, the major challenge of a live cell DEL assay is the low fractional binding of a DEL hit to the target at equilibrium due to the low protein concentration and modest affinity of hit ligands. Any single compound within a diverse DEL is present at an extremely low concentration (typically the fM–pM range), which is likely well below the K_d for the target. Thus, DEL assays rely on the target protein concentration to drive the binding equilibrium. Several groups have addressed this challenge with various approaches: high-level overexpression, covalent crosslinking, bivalent ligands, label transfer of ssDNA tags, and enzymatic proximity labeling.

Not surprisingly, most activity in this area has been with cell surface proteins. These proteins comprise a large class of drug targets and are challenging to use in the traditional DEL assay of an affinity selection with a purified protein. We anticipate refinement of these techniques will make live cell surface selections more mainstream. The ability to fully capitalize on the massive chemical complexity of DELs should provide major benefits to drug development for these targets. Improvements in agonist specific or other activity-specific selections should additionally provide benefits. While several demonstrations to date have been successful, it is clear that cell surface selections yield much "noisier" results than analogous selections with a pure protein target. Liu and coworkers have applied machine learning (ML) models to improve interpretation of these data.[39]

The application of DELs to protein targets within live cells has seen less application. A major additional challenge here is the impermeability of small molecule–DNA conjugates. Initial reports using cell penetrating peptides with mammalian cells and microinjection into *Xenopus* oocytes have provided promising demonstrations. Indeed, for many targets, there may not be a strong justification for pursuing a live cell selection. As methods for assaying DELs against targets within live cells improve, the ability to target proteins that are inaccessible to biochemical purification can offer some exciting possibilities.

A major justification for pursuing live cell selections is the ability to probe the target protein in its physiologically relevant state. Developed methods to date vary in how relevant that state may be. Some approaches enable the targeting of endogenous proteins. Other approaches require recombinant expression with fusion protein tags in human cells or *Xenopus* oocytes. The requirements for obtaining a protein in the desired active state will certainly vary from target to target. All the presented approaches remove the requirement of upfront purification of the target protein, which can be a major benefit over the conventional approach. Requirements for characterization of the protein activity level are also greatly reduced or eliminated. In many cases, obtaining a completely pure protein and determining its specific activity can be an onerous part of a DEL drug discovery campaign.

As the use of DELs in drug discovery continues to increase, we anticipate that there will be many more applications of live cell selections. Further technology development and refinement will facilitate greater adoption. For example, recent developments with enzymatic proximity labeling with fusion proteins promise to facilitate the detection of live cell targets under low protein concentration conditions through signal amplification.[23,40] Additional assay technology promises to expand the frontiers of what is possible for DELs in drug discovery.

Acknowledgements

This work was supported by a grant from the National Institutes of Health to C. J. K. (R35GM128894). Figures 4.1 and 4.3 were created with BioRender.com.

References

1. T. L. Foley, W. Burchett, Q. Chen, M. E. Flanagan, B. Kapinos, X. Li, J. I. Montgomery, A. S. Ratnayake, H. Zhu and M.-C. Peakman, *SLAS Discovery*, 2021, **26**, 263–280.

2. A. L. Satz, A. Brunschweiger, M. E. Flanagan, A. Gloger, N. J. V. Hansen, L. Kuai, V. B. K. Kunig, X. Lu, D. Madsen and L. A. Marcaurelle, *Nat. Rev. Methods Primers*, 2022, **2**, 3.
3. D. R. Halpin, J. A. Lee, S. J. Wrenn and P. B. Harbury, *PLoS Biol.*, 2004, **2**, E175.
4. Y. Shi, Y. Wu, J. Yu, W. Zhang and C. Zhuang, *RSC Adv.*, 2021, **11**, 2359–2376.
5. B. Briard, D. E. Place and T.-D. Kanneganti, *Physiology*, 2020, **35**, 112–124.
6. K. Sriram and P. A. Insel, *Mol. Pharmacol.*, 2018, **93**, 251–258.
7. D. G. Brown, G. A. Brown, P. Centrella, K. Certel, R. M. Cooke, J. W. Cuozzo, N. Dekker, C. E. Dumelin, A. Ferguson and C. Fiez-Vandal, *et al.*, *SLAS Discovery*, 2018, **23**, 429–436.
8. S. Ahn, A. W. Kahsai, B. Pani, Q.-T. Wang, S. Zhao, A. L. Wall, R. T. Strachan, D. P. Staus, L. M. Wingler and L. D. Sun, *Proc. Natl. Acad. Sci. U. S. A.*, 2017, **114**, 1708–1713.
9. S. Ahn, B. Pani, A. W. Kahsai, E. K. Olsen, G. Husemoen, M. Vestergaard, L. Jin, S. Zhao, L. M. Wingler and P. K. Rambarat, *et al.*, *Mol. Pharmacol.*, 2018, **94**, 850–861.
10. Z. Wu, T. L. Graybill, X. Zeng, M. Platchek, J. Zhang, V. Q. Bodmer, D. D. Wisnoski, J. Deng, F. T. Coppo, G. Yao, A. Tamburino, G. Scavello, G. J. Franklin, S. Mataruse, K. L. Bedard, Y. Ding, J. Chai, J. Summerfield, P. A. Centrella, J. A. Messer, A. J. Pope and D. I. Israel, *ACS Comb. Sci.*, 2015, **17**, 722–731.
11. Y. Gui, R. Hou, Y. Huang, Y. Zhou, S. Liu, L. Meng, Y. Li, F. S. Lam, C. S. Wong, C. T. Yin, G. Li, X. Lu and X. Li, *ChemRxiv*, 2024, DOI: 10.26434/chemrxiv-2024-3mrz3.
12. B. Cai, A. El Daibani, Y. Bai, T. Che and C. J. Krusemark, *JACS Au*, 2023, **3**, 1076–1088.
13. Y. Huang, L. Meng, Q. Nie, Y. Zhou, L. Chen, S. Yang, Y. M. E. Fung, X. Li, C. Huang, Y. Cao, Y. Li and X. Li, *Nat. Chem.*, 2021, **13**, 77–88.
14. L. K. Petersen, A. B. Christensen, J. Andersen, C. G. Folkesson, O. Kristensen, C. Andersen, A. Alzu, F. A. Sløk, P. Blakskjær, D. Madsen, C. Azevedo, I. Micco and N. J. V. Hansen, *J. Am. Chem. Soc.*, 2021, **143**, 2751–2756.
15. M. I. Love, W. Huber and S. Anders, *Genome Biol.*, 2014, **15**, 1–21.
16. B. Cai, D. Kim, S. Akhand, Y. Sun, R. J. Cassell, A. Alpsoy, E. C. Dykhuizen, R. M. Van Rijn, M. K. Wendt and C. J. Krusemark, *J. Am. Chem. Soc.*, 2019, **141**, 17057–17061.
17. H. Xiong, J. Yan, S. Cai, Q. He, D. Peng, Z. Liu and Y. Liu, *Int. J. Biol. Macromol.*, 2019, **132**, 190–202.
18. B. Shi, Y. Deng, P. Zhao and X. Li, *Bioconjugate Chem.*, 2017, **28**, 2293–2301.
19. K. E. Denton and C. J. Krusemark, *MedChemComm*, 2016, **7**, 2020–2027.
20. G. Li, Y. Liu, Y. Liu, L. Chen, S. Wu, Y. Liu and X. Li, *Angew. Chem., Int. Ed.*, 2013, **52**, 9544–9549.
21. K. Kolberg, C. Puettmann, A. Pardo, J. Fitting and S. Barth, *Curr. Pharm. Des.*, 2013, **19**, 5406–5413.

22. Y. Gui, C. S. Wong, G. Zhao, C. Xie, R. Hou, Y. Li, G. Li and X. Li, *ACS Omega*, 2022, **7**, 11491–11500.
23. B. Cai, A. B. Mhetre and C. J. Krusemark, *Chem. Sci.*, 2023, **14**, 245–250.
24. L. Kubitz, S. Bitsch, X. Zhao, K. Schmitt, L. Deweid, A. Roehrig, E. C. Barazzone, O. Valerius, H. Kolmar and J. Béthune, *Commun. Biol.*, 2022, **5**, 657.
25. Q. Zhang, Y. Zheng, S. R. Coughlin and X. Shu, *Protein Sci.*, 2018, **27**, 874–879.
26. M. Spetea, S. O. Eans, M. L. Ganno, A. Lantero, M. Mairegger, L. Toll, H. Schmidhammer and J. P. McLaughlin, *Br. J. Pharmacol.*, 2017, **174**, 2444–2456.
27. A. Cannaert, L. Vasudevan, M. Friscia, A. L. A. Mohr, S. M. R. Wille and C. P. Stove, *Clin. Chem.*, 2018, **64**, 1221–1229.
28. K. F. Cho, T. C. Branon, S. Rajeev, T. Svinkina, N. D. Udeshi, T. Thoudam, C. Kwak, H.-W. Rhee, I.-K. Lee, S. A. Carr and A. Y. Ting, *Proc. Natl. Acad. Sci. U. S. A.*, 2020, **117**, 12143–12154.
29. I. Gomes, S. Sierra, L. Lueptow, A. Gupta, S. Gouty, E. B. Margolis, B. M. Cox and L. A. Devi, *Proc. Natl. Acad. Sci. U. S. A.*, 2020, **117**, 11820–11828.
30. W. Huang, A. Manglik, A. J. Venkatakrishnan, T. Laeremans, E. N. Feinberg, A. L. Sanborn, H. E. Kato, K. E. Livingston, T. S. Thorsen and R. C. Kling, *Nature*, 2015, **524**, 315–321.
31. H.-W. Rhee, P. Zou, N. D. Udeshi, J. D. Martell, V. K. Mootha, S. A. Carr and A. Y. Ting, *Science*, 2013, **339**, 1328–1331.
32. S. Oehler, M. Catalano, I. Scapozza, M. Bigatti, G. Bassi, N. Favalli, M. R. Mortensen, F. Samain, J. Scheuermann and D. Neri, *Chem. – Eur. J.*, 2021, **27**, 8985–8993.
33. Z. Qian, A. Martyna, R. L. Hard, J. Wang, G. Appiah-Kubi, C. Coss, M. A. Phelps, J. S. Rossman and D. Pei, *Biochemistry*, 2016, **55**, 2601–2612.
34. L. Peraro, Z. Zou, K. M. Makwana, A. E. Cummings, H. L. Ball, H. Yu, Y. S. Lin, B. Levine and J. A. Kritzer, *J. Am. Chem. Soc.*, 2017, **139**, 7792–7802.
35. J. S. Marchant, *Cold Spring Harb. Protoc.*, 2018, pdb-prot096990.
36. P. Blakskjaer, T. Heitner and N. J. V. Hansen, *Curr. Opin. Chem. Biol.*, 2015, **26**, 62–71.
37. Vipergen: Cellular Binder Trap Enrichment, https://www.vipergen.com/cellular-binder-trap-enrichment/, (Accessed June 2024).
38. S. Wang, K. E. Denton, K. F. Hobbs, T. Weaver, J. M. B. McFarlane, K. E. Connelly, M. C. Gignac, N. Milosevich, F. Hof, I. Paci, C. A. Musselman, E. C. Dykhuizen and C. J. Krusemark, *ACS Chem. Biol.*, 2020, **15**, 112–131.
39. R. Hou, C. Xie, Y. Gui, G. Li and X. Li, *ACS Omega*, 2023, **8**, 19057–19071.
40. L. A. Schneider, B. Sauter, K. Dagher and D. Gillingham, *J. Am. Chem. Soc.*, 2023, **145**, 20874–20882.
41. J. Deng, S. Belyanskaya, N. Prabhu, C. Arico-Muendel, H. Deng, C. B. Phelps, D. I. Israel, H. Yang, J. Boyer and G. J. Franklin, *SLAS Discovery*, 2024, 100171.

CHAPTER 5

On-DNA Medicinal Chemistry: Focused Libraries, Fragment Expansion, and Hit Optimisation

H. A. STANWAY-GORDON AND M. J. WARING*

Cancer Research Horizons Therapeutic Innovation Newcastle Drug Discovery Group, Chemistry, School of Natural and Environmental Sciences, Newcastle University, Bedson Building, UK
*Email: mike.waring@ncl.ac.uk

5.1 Introduction

From their inception by Brenner and Lerner in 1992, DNA-encoded libraries (DELs) were primarily envisioned to be utilised as a medium to facilitate hit finding.[1] Much of the subsequent published work has indeed demonstrated that DELs can be successfully employed for this purpose, with application of synthetically constructed encoded-libraries resulting in the generation of lead-series across a wide variety of target classes.[2–7] In recent years however, multiple groups have demonstrated that the utility of DELs within drug discovery is not necessarily limited to this primary aim and a variety of strategies to broaden the impact of encoded libraries within the field have been adopted.

In this chapter, we discuss the current applications of DELs to the wider medicinal chemistry pipeline, in particular exploring the application and impact of focused libraries, fragment discovery and expansion with encoded

Drug Discovery Series No. 85
DNA-encoded Library Technology for Drug Discovery
Edited by Guansai Liu, Casey J. Krusemark and Jin Li

Published by the Royal Society of Chemistry, www.rsc.org

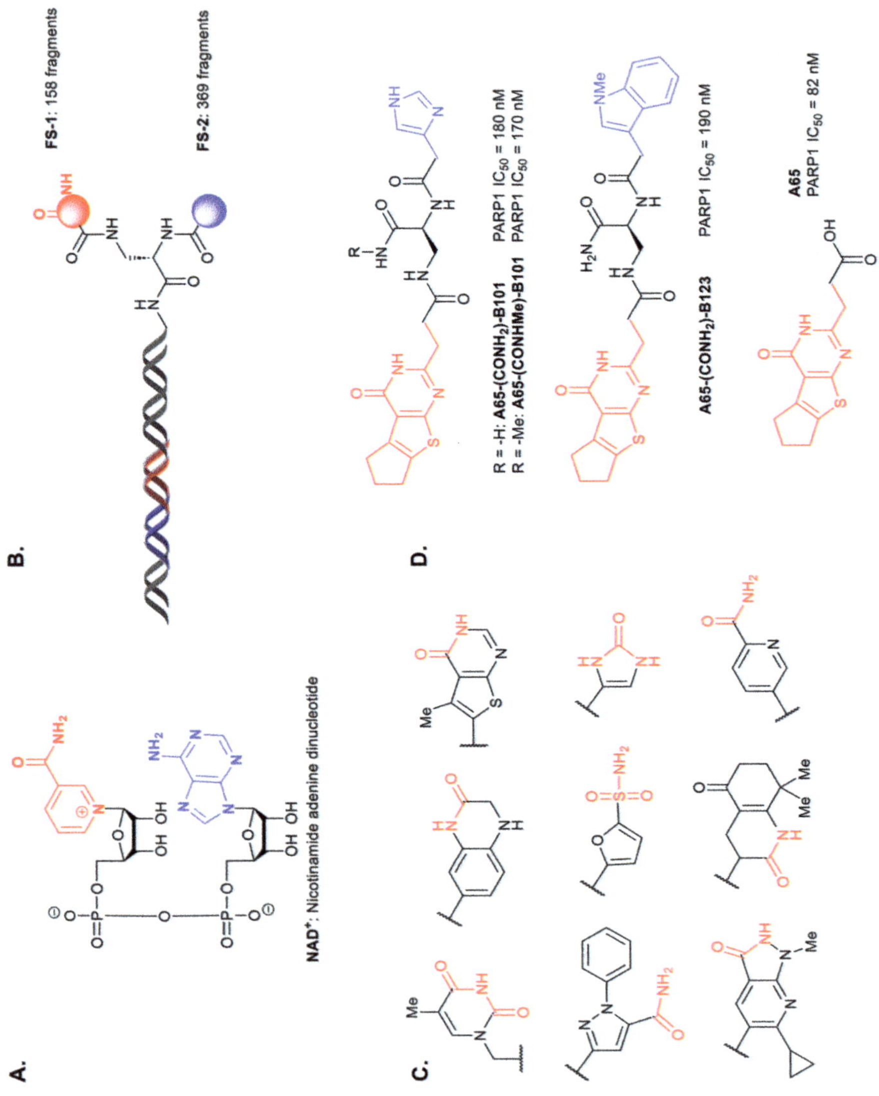
A.
NAD$^+$: Nicotinamide adenine dinucleotide
B.
FS-1: 158 fragments
FS-2: 369 fragments
C.
D.
R = -H: A65-(CONH$_2$)-B101 PARP1 IC$_{50}$ = 180 nM
R = -Me: A65-(CONHMe)-B101 PARP1 IC$_{50}$ = 170 nM
A65-(CONH$_2$)-B123 PARP1 IC$_{50}$ = 190 nM
A65
PARP1 IC$_{50}$ = 82 nM

libraries, alongside hit optimisation strategies that have been performed on-DNA. We assess the current state of the field within this role and potential future directions for the technology within this area.

5.2 Focused Libraries

Typically when DELs have been employed in hit-finding campaigns, one of the key elements of their design has been to encompass as wide a range of chemical space as possible with the understanding that such a feature will lead to a higher hit discovery rate.[8–10] In contrast to this convention, focused or targeted libraries are those that are designed with a specific target in mind and chemically biased accordingly. Such libraries may be centred around the use of a privileged scaffold, incorporate specific recognition elements for a protein of interest, be designed around an architecture representative of an established ligand, or possess an over-representation of classes of building blocks reported to preferentially interact with the desired target.

Franzini *et al.* were one of the first groups to report a structurally focused DEL for the discovery of inhibitors of NAD^+ dependent enzymes in 2019.[11] Designing a library with a general compound structure mimicking the natural substrate, a 58 302-member library, termed NADEL, was generated employing 527 fragments (158 × 369) arranged in a branched geometry on a 2,3-diaminopropionamide scaffold (Figure 5.1). Fragments at FS-1 were selected to contain a hydrogen donor/acceptor pattern akin to the carboxamide group of NAD^+, whilst FS-2 was unbiased. Initially selected against PARP1, results demonstrated strong enrichment for library members with a cyclopentathienopyrimidinone fragment (A65) at FS-1. Resynthesis of the two most enriched structures was performed, in conjunction with fragment A65 with all possessing IC_{50} values in the 100 nM region; several structures were also demonstrated to inhibit the closely related isoform PARP2 ($IC_{50} < 50$ nM). To assess the general utility of NADEL, multiple further selections were performed across the PARP family and SIRT6, yielding the identification of numerous ligands with a broad range of potencies ($IC_{50} = 0.2$–38 μM).[11] When comparing the results of NADEL selections to those performed using an unbiased structurally related library (DAL-100K) constructed utilising the same chemistry, significantly fewer hits were identified exemplifying the focused-library approach within the field and demonstrating the potential of this method.

Figure 5.1 Structurally focused NADEL reported by Franzini *et al.* (A) Structure of NAD^+ on which the design of NADEL was based. (B) General structure of NADEL containing two points of diversity. FS-1 consists of 158 fragments containing a H-donor/acceptor pattern akin to the carboxamide group of NAD^+. FS-2 contains 369 unbiased fragments. (C) Exemplary structures of building blocks incorporated at FS-1 with carboxamide mimetics highlighted. (D) Validated hits identified from the selection of NADEL against PARP1.[11] Adapted from ref. 11 with permission from American Chemical Society, Copyright 2019.

Matzuk *et al.* used a similar approach to identify thrombin inhibitors from a protease-focused DEL using argatroban, a licensed thrombin inhibitor as a template for the library design. Incorporating known functional groups of interest, *e.g.*, guanidine, sulfonamide, and carbamate, the group synthesised a 3-cycle 9.8-million-member library *via* sequential amide coupling/deprotection chemistries terminating with an amine capping step (Figure 5.2).[12] Selection against thrombin displayed strong structure–enrichment relationships for a number of trisynthons and two structurally related compounds were selected for off-DNA validation, differing solely in the identity of BB3. Initial assessment of the isolated structures was performed on-DNA, with inhibitory potencies observed to be 58 and 589 nM for compounds **5** and **6**, respectively. Compound **5** was subsequently synthesised off-DNA and demonstrated to have an inhibitory potency of 1 nM, *ca.* 10× more potent than the library template argatroban ($K_i = 9$ nM).

Pearce *et al.* employed a focused library approach to identify small molecule antagonists for methyl-lysine (KMe) reader p53 binding protein 1 (53BP1).[13] Compounds within the library were designed to be bivalent binders with a bias towards KMe-reader proteins containing three target-relevant fragments: BB1 consisted of KMe-cage binding fragments, BB3 was intended to interact with a second site of interest within the targets (*e.g.* a secondary KMe/KAc binding site or hydrophobic patch), and BB2 acted as a linker (Figure 5.3).[14]

Initial selections performed with the library against a myriad of epigenetic targets (53BP1, KDM7B, CDYL2, CBX2 and LEDGF) led to the identification of μM hits obtained for 53BP1 and KDM7B; no compounds of interest were identified in the remaining cases.[15] Subsequent publication of a replicate selection of the same library against 53BP1 demonstrated that more potent hits could be identified, with the initial off-DNA validation of enriched compound UNC8531 possessing sub-μM potency in multiple orthogonal assays. Further optimisation derived from co-crystallisation and subsequent SAR studies yielded UNC9512, a best-in-class 53BP1 TTD antagonist ($K_d = 170$ nM).[13]

Zhao *et al.* reported the synthesis and selection of a phenolic acid focused library for the identification of angiotensin II type I receptor (AT_1R) antagonists in 2021 (Figure 5.4).[16] Using split-and-pool chemistry, the group constructed a 32 000-member library containing three cycles of amino acids subsequently terminated by one of four phenolic acids. Selection of the library yielded seven encoded compounds exhibiting significant enrichment (>20× no target control, NTC). The most enriched compound had a K_d of 0.49 nM.

A targeted approach has also been utilised for the discovery of ligands for the RNA repeat expansion $r(CUG)^{exp}$. Disney *et al.* synthesised a solid-phase 3-cycle 12 672-member DEL designed to be enriched in moieties common in RNA binders including nitrogen rich heterocycles and increased numbers of hydrogen bond donors/acceptors (Figure 5.5).[17] Employing a fluorescence-based assay, the library was selected against Cy5-$r(CUG)_{12}$ and Alexa750-$r(GC)_8$,

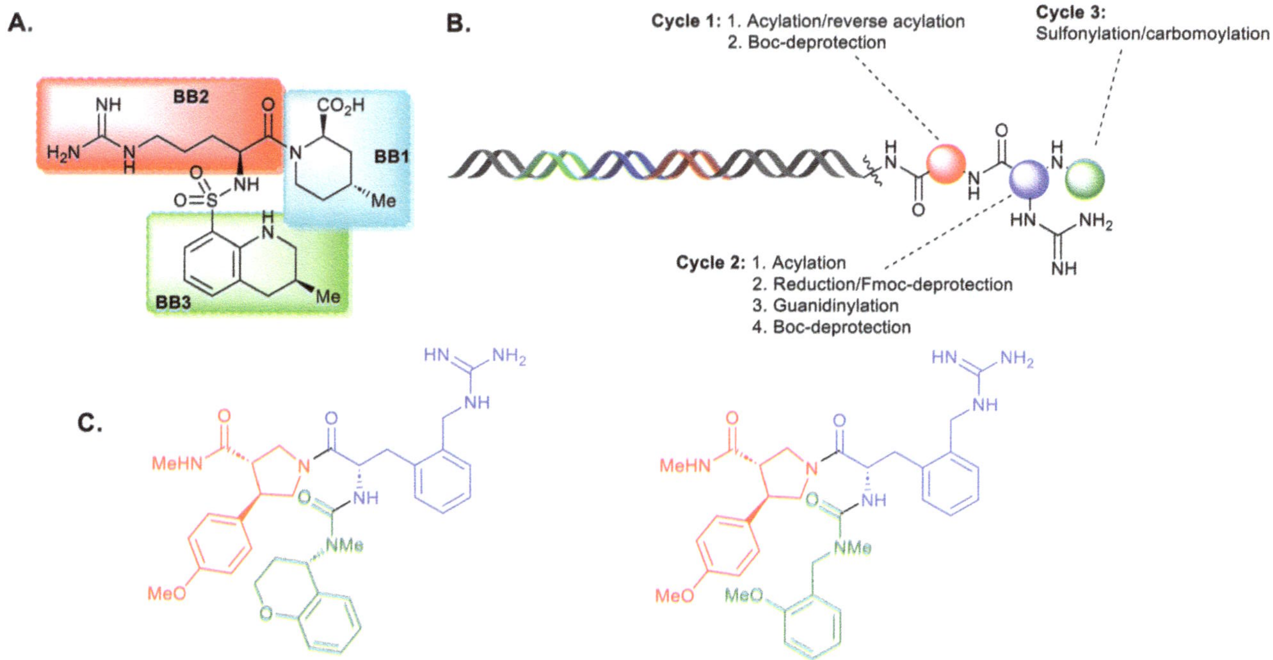

DNA-5 thrombin $K_i$app = 58 nM
5 thrombin $K_i$app = 1 nM

DNA-6 thrombin $K_i$app = 589 nM

Figure 5.2 Protease-focused DEL utilised for the identification of thrombin inhibitors. (A) Structure of argatroban which was used to inform the design of the library with suggested mapping of points of diversity by the authors. (B) Protease focused DEL structure with the indication of reactions within each cycle. (C) Most enriched compounds identified from the selection of the library against thrombin and measured inhibitory potencies of isolated oligonucleotide conjugated derivatives or off-DNA equivalents as indicated.[12] Adapted from ref. 12, https://doi.org/10.1073/pnas.2005447117, under the terms of the CC BY 4.0 license, https://creativecommons.org/licenses/by/4.0/.

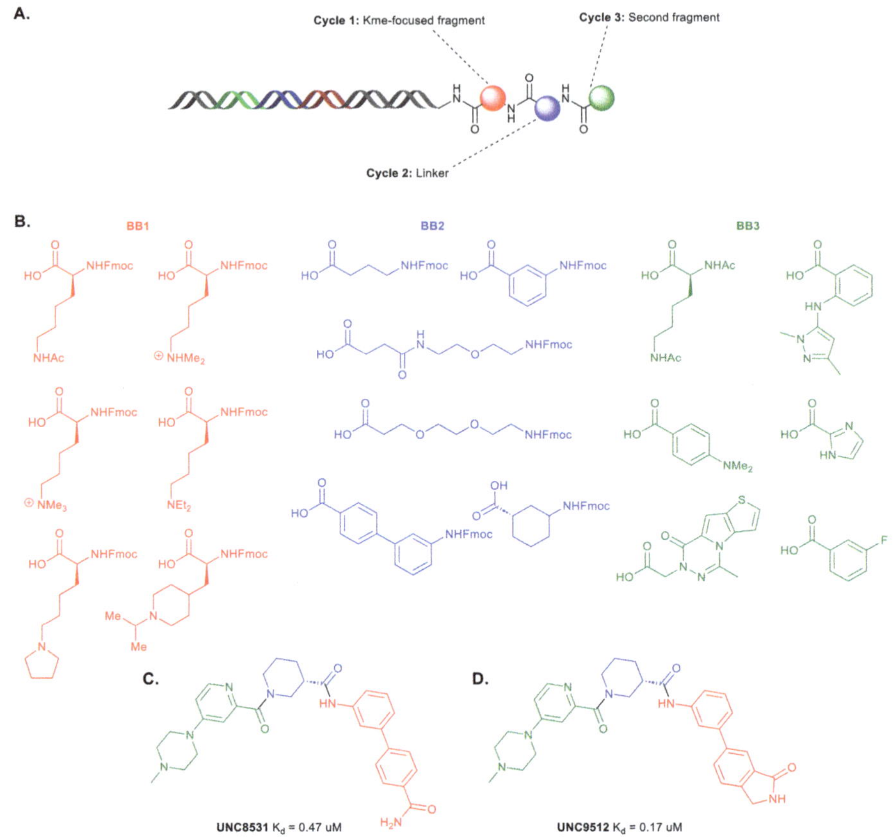

Figure 5.3 Focused library reported by Pearce *et al.* for application to epigenetic targets. (A) General library design and relevance of each point of diversity within the library. (B) Exemplary building blocks employed for each of the respective positions. (C) UNC8531, hit identified from selection of the library against 53BP1. (D) UNC9512, optimised hit resulting from SAR investigation of UNC8531.[13–15] Adapted from ref. 14, https://doi.org/10.3390/molecules25040979, under the terms of the CC BY 4.0 license, https://creativecommons.org/licenses/by/4.0/.

ultimately leading to 23 hit compounds selective towards the target. Of these, 16 were followed up in a phenotypic assay with a single compound showing significantly improved activity over the others selected ($K_d = 6.3$ μM), demonstrating the applicability of focused libraries to more unconventional targets.

Paegel *et al.* reported the use of a "property-focused" DEL to explore potential therapeutics for multidrug-resistant (MDR) pathogens (Figure 5.6).[18] Designing a library centred around the physicochemical property space of established antibacterials, the group constructed a 7488-member three-cycle solid-phase DEL, with a focus on reduced hydrophobicity and higher molecular weight. In contrast to conventional methods, the group did not employ affinity selection for the identification of hits, but instead developed

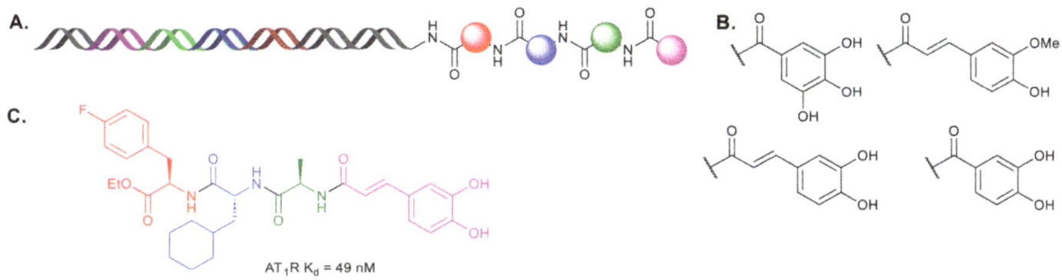

Figure 5.4 Phenolic acid-based library reported by Zhao *et al.* for the identification of AT_1R antagonists.[16] (A) General structure of the library synthesised *via* three cycles of coupling with amino acids prior to focussing with phenolic acids. (B) Structures of capping phenolic acids incorporated within the library. (C) Most enriched structure identified from DEL selection. Adapted from ref. 16 with permission from American Chemical Society, Copyright 2021.

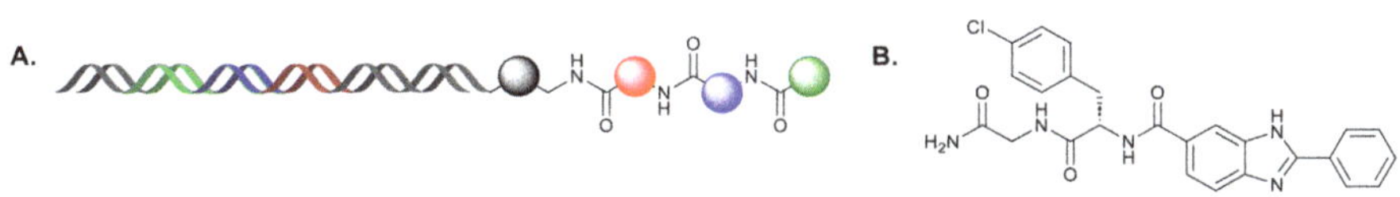

Figure 5.5 Solid-phase library employed by Disney *et al.* for the discovery of RNA binding ligands.[17] (A) General architecture of the library constructed *via* sequential amide couplings. (B) Structure of the most enriched compound from selection against $r(CUG)^{exp}$. Adapted from ref. 17 with permission from American Chemical Society, Copyright 2022.

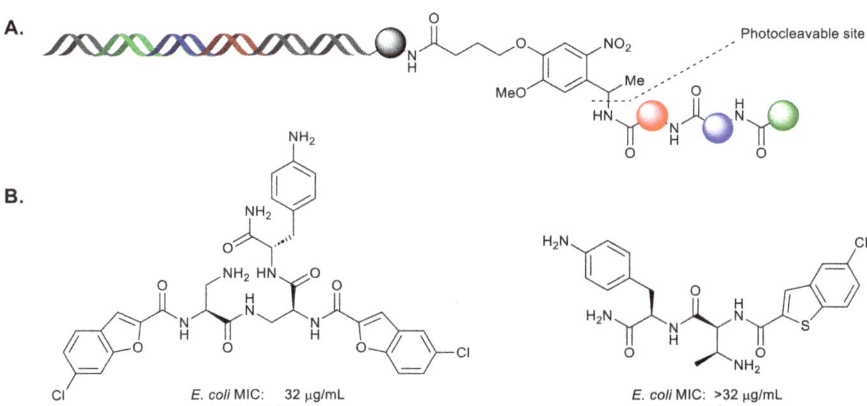

Figure 5.6 Property-focused DEL employed for exploration of potential therapeutics for MDR pathogens.[18,19] (A) General structure of the encoded library including the photocleavable linker required for selection. (B) Exemplary hit compounds identified with activity against *Bacillus subtilis*. Adapted from ref. 19 https://doi.org/10.1021/acsmedchemlett.3c00159, under the terms of the CC BY 4.0 license, https://creativecommons.org/licenses/by/4.0/.

a novel phenotypic approach using a whole cell bead diffusion assay.[19] Containing a photocleavable linker, UV irradiation of the beads allowed for release of compounds and subsequent diffusion into the surrounding media; hit compounds accordingly resulted in a growth inhibition zone (GIZ) of the target bacteria after incubation.

97 hit beads were identified from a total of 85 000 screened, with two micromolar inhibitors of *Bacillus subtilis* identified and selection outputs related to fluoroquinolone containing members correlating with the established SAR around the class. Of note, the structure of several hits identified from the selection were determined to be unexpected side-products during validation synthesis and not the expected products. This exemplifies the importance of the resynthesis and testing of compounds identified in DEL selections before they can be considered as hits.

Targeted protein degradation and the use of proteolysis-targeting chimeras (PROTACs) as a therapeutic modality is a rapidly expanding area of interest in drug discovery, with the potential to pursue previously deemed "undruggable" targets. In 2019, the first PROTACs entered clinical trials, with clinical proof-of-concept data being reported the subsequent year. Accordingly, there is significant interest in the application and development of such molecules.[20]

Schreiber *et al.* reported the successful use of DELs to facilitate the direct discovery of chemical inducers of proximity (CIPs) in 2024.[21] Using $BRD4^{BD1}$ as a test system, the group designed a 1-million-member library (CIP-DEL) containing three points of diversity based on the structure of MZ1, an established degrader of the target (Figure 5.7). Affinity selections were performed against immobilised BRD4 in both the presence and the absence of VHL to identify promising CIPs from the library. Validation was performed with multiple hit compounds in addition to several non-enriched members to assess the suitability of the assay. A number of highly potent CIPs were identified, with affinity to both BRD4 and VHL measured independently prior to assessment of ternary complex formation and protein degradation. The group also reported additional selections against the first bromodomains of BRD2 and BRDT.

Most recently, Palzkill *et al.* reported the use of a focused DEL to deduce hits containing novel pharmacophores against β-lactamases in efforts to identify co-treatments for antimicrobial resistance (AMR).[22] Targeting the highly conserved carboxylate binding pocket of the enzyme family, the group synthesised a number of libraries with different chemistries (Figure 5.8), terminating all with a carboxylic acid- or tetrazole-containing building block to act as a recognition element. 228 750 members of the focused DELs were selected in conjunction with 12 additional libraries (total 6.28 million members) against OXA-48 (a class D serine hydrolase) and NDM-1 (a class B zinc metallo-enzyme). For OXA-48, several hits were identified, with four compounds from the focused library demonstrating activity in an enzyme inhibition assay ($K_i = 0.17$ to 130 μM).[22] For NDM-1, of the 13 libraries screened, only the focused library yielded any hits, with the most potent

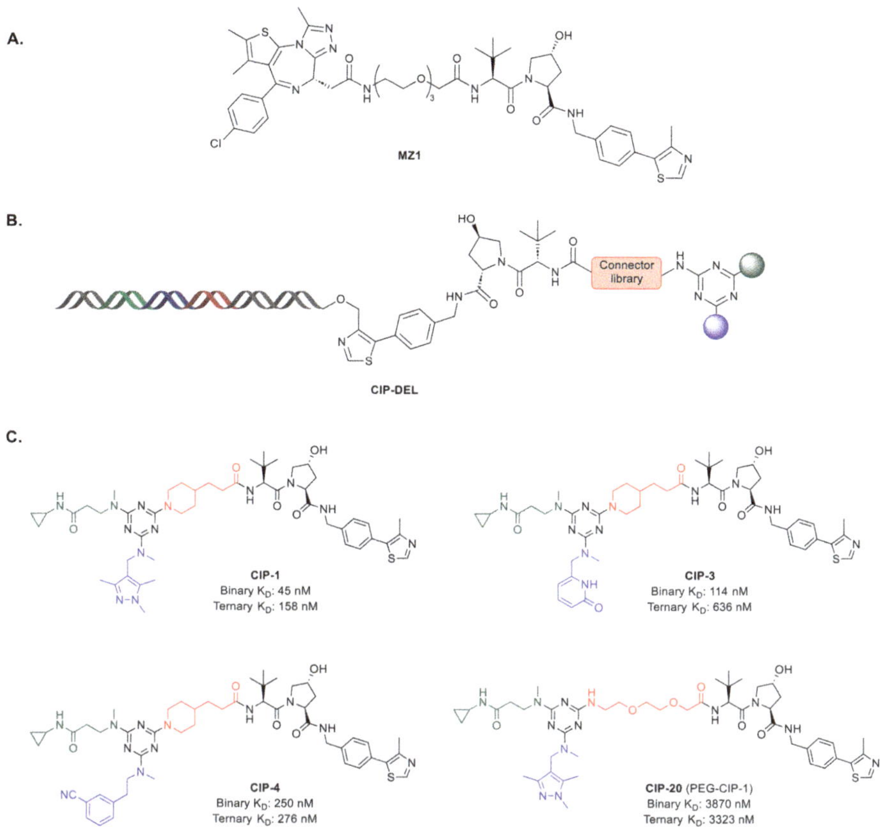

Figure 5.7 Overview of the design and selection of CIP-DEL. (A) Structure of MZ-1, an established degrader of BRD4 from which library design was based. (B) General structure of CIP-DEL. (C) Synthesised compounds following selection of the library against BRD4 and VHL. CIP-1/2/4 demonstrated significant enrichment in the assay and were confirmed as hits following off-DNA synthesis. CIP-20 did not demonstrate enrichment and subsequent synthesis was used to validate the efficiency of the selection process.[21] Adapted from ref. 21 with permission from Springer Nature, Copyright 2024.

compound demonstrating a K_i of 3.3 μM. Further off-DNA optimisation of this structure identified an improved inhibitor ($K_i = 0.6$ μM). The authors state that no false positive hits were identified during the selections, suggesting that a focused library approach has the potential to offer a significant improvement over conventional DEL-screening.[23]

In addition to the methods discussed previously, several groups have reported strategies for the synthesis of scaffold focused DELs. In contrast to the target-based methods above, such approaches are centred around the ability to synthesise certain privileged scaffolds on-DNA, which are anticipated to result in a higher hit-rate due to the prevalence of these structures

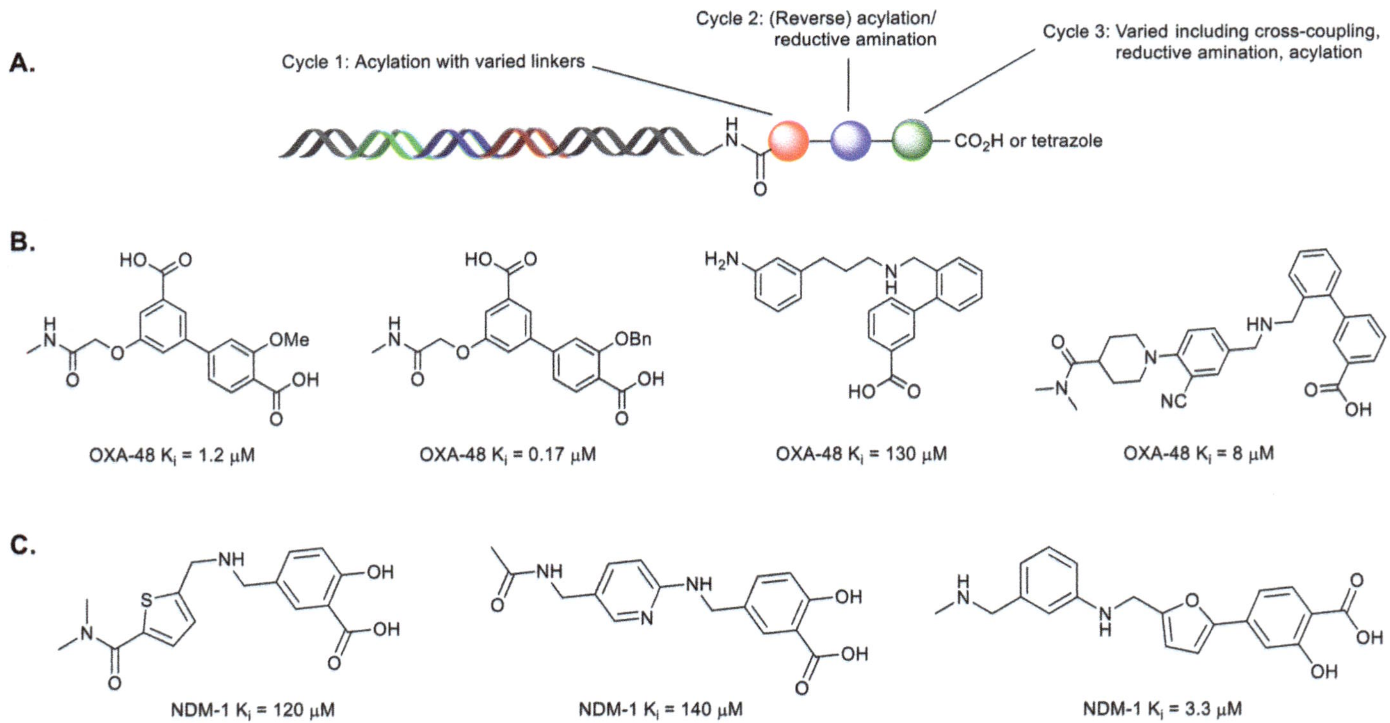

Figure 5.8 Application of a pharmacophore-focused DEL for identification of β-lactamase inhibitors.[22] (A) General structure of the library. (B) Hits identified from selection against OXA-48. (C) Hits obtained from selection against NDM-1. Adapted from ref. 22 with permission from American Chemical Society, Copyright 2024.

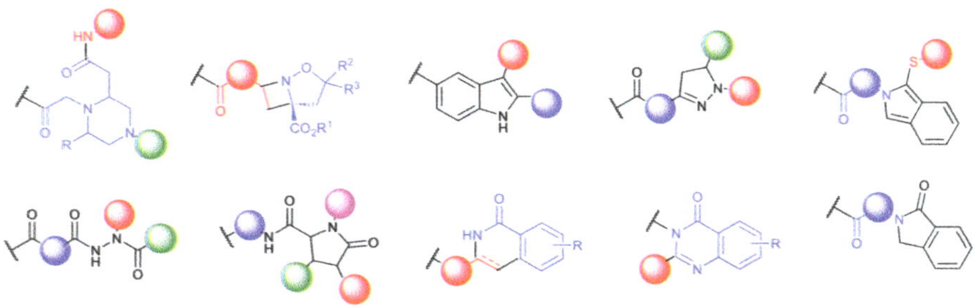

Figure 5.9 Examples of reported accessible structures within DELs for the generation of scaffold focused libraries indicating potential points of diversity within library synthesis and representative points of attachment to DNA.[24–43]

within established drugs.[24–43] Such a hypothesis currently remains to be realised based on the available results within the literature; whilst some examples have been incorporated into full-size libraries with hits identified, the majority are either only disclosures of reaction development and/or no subsequent hits have been disclosed from the reported synthesised libraries (Figure 5.9).

5.3 Fragments and Fragment Expansion

Fragment-based drug discovery (FBDD) has been an important avenue in medicinal chemistry over the past 30 years with a number of successful campaigns leading to both approved drugs and multiple clinical candidates.[46] FBDD can be an incredibly powerful technique with regard to the range of chemical space that can be covered within the libraries and accordingly the hit rate observed; however, due to the fewer interactions formed by molecules being screened as a result of their reduced molecular weight, library members have a lower affinity for targets of interest and as such assay sensitivity is of great importance. The identification of fragments from DELs *via* typical affinity selection would therefore be impractical, as structures of interest would not be sufficiently retained in DEL selection procedures.

Whilst previously utilised within DELs for the identification of fully elaborated hit compounds,[47] Liu *et al.* were the first group to demonstrate the capabilities of photo-crosslinking in the application of DELs to FBDD, termed PAC-FragmentDEL. The group synthesised a series of sub-libraries *via* split-and-pool methodology, all containing a diazirine linker between the compounds and the encoding oligonucleotide (Figure 5.10).[48]

The sub-libraries could be classified into two types: the first contained compounds with properties observed in conventional fragment libraries, with the majority described as having <18 heavy atoms, a clog $P<2$, <3 H-bond acceptors and 2 H-bond donors. The latter consisted of merged fragments that, whilst outside of this desired space, the authors

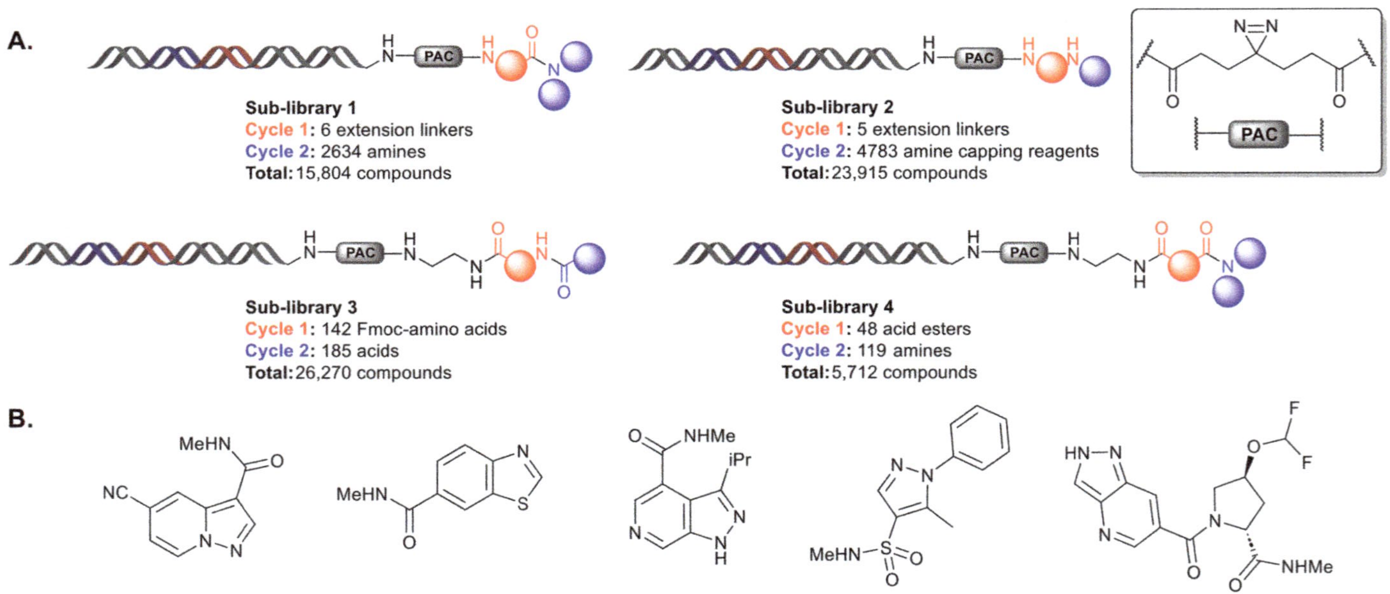

Figure 5.10 Overview of PAC-FragmentDEL for identification of encoded fragment hits. (A) General structure of sub-libraries contained within PAC-FragmentDEL. (B) Hits identified from the selection of PAC-FragmentDEL against 2-epimerase. Adapted from ref. 48 with permission from the Royal Society of Chemistry.[48]

theorised would be beneficial for screening binding sites with lower ligand efficiencies.

Employing PAK4 as a proof-of-concept target, the group successfully demonstrated that established encoded fragments could be sufficiently enriched *via* the selection method in contrast to control experiments in the absence of protein and/or UV light. Subsequent selection of the fragment libraries against PAK4 identified 301 fragments of interest; 11 of these were subjected to further validation and binding was confirmed in all cases *via* ligand observed NMR (LO-NMR), with crystal structures also obtained in 10 cases. PAC-FragmentDEL was also selected against 2-epimerase; however, in contrast to PAK4, significantly less enrichment was observed, indicative of the lower ligandability of this target. 21 competitive fragments were identified, with 9 selected for validation; again, LO-NMR confirmed binding in all instances and crystal structures were obtained for 5 of the hits.

The widespread application of DELs for FBDD remains to be determined; however, significantly more work has been performed on DELs employed for fragment expansion. Such efforts typically involve the incorporation of a warhead or consistent fragment of interest within the library, with encoded components used to expand upon this, generating fully elaborated compounds as the resulting hits.

Whilst sequence recorded split-and-pool has become the dominant technique for DEL synthesis, for fragment expansion, encoded self-assembling compound libraries (ESAC) have shown significant promise due to their unique method of construction.[49] ESAC technology relies on the hybridisation of oligonucleotides to construct the libraries, with small molecules conjugated to either the 5′ or 3′ ends of partially complementary encoding strands. Hybridisation of the oligonucleotides yields pairs of small molecules that are selected together; depending on the sequences involved in the complementary regions of the DNA, these libraries can be either static or dynamic, with both approaches having been explored, alongside alternative oligonucleotide constructs for library generation.[50,51] There have been numerous reports of the presence of a sole established ligand on one strand, with the variation of fragments attached to the partially complementary partner. In this manner, the original ligand can be further elaborated by synthetic conjugation with the paired molecule that is identified as enriched by selection, *via* a process termed "affinity maturation" (Figure 5.11).[49,52–56] More recent examples of fragment expansion *via* the medium of ESAC are discussed herein.

In 2020, Neri *et al.* reported the use of a dual-pharmacophore library against the cyclic-AMP response element binding protein (CREBBP) bromodomain.[57] An established 4,5-dihydrobenzodiazepinone (THBD)-based ligand was employed as the known fragment and encoded within two sublibraries consisting of 787 and 424 members, respectively (total library size of *ca.* 335k). Selection against the target yielded a pair of enriched fragments containing the (*R*)-THBD moiety in conjunction with the dithiobenzoate derivative A296 (Figure 5.12). Hit-validation was performed on-DNA, with the

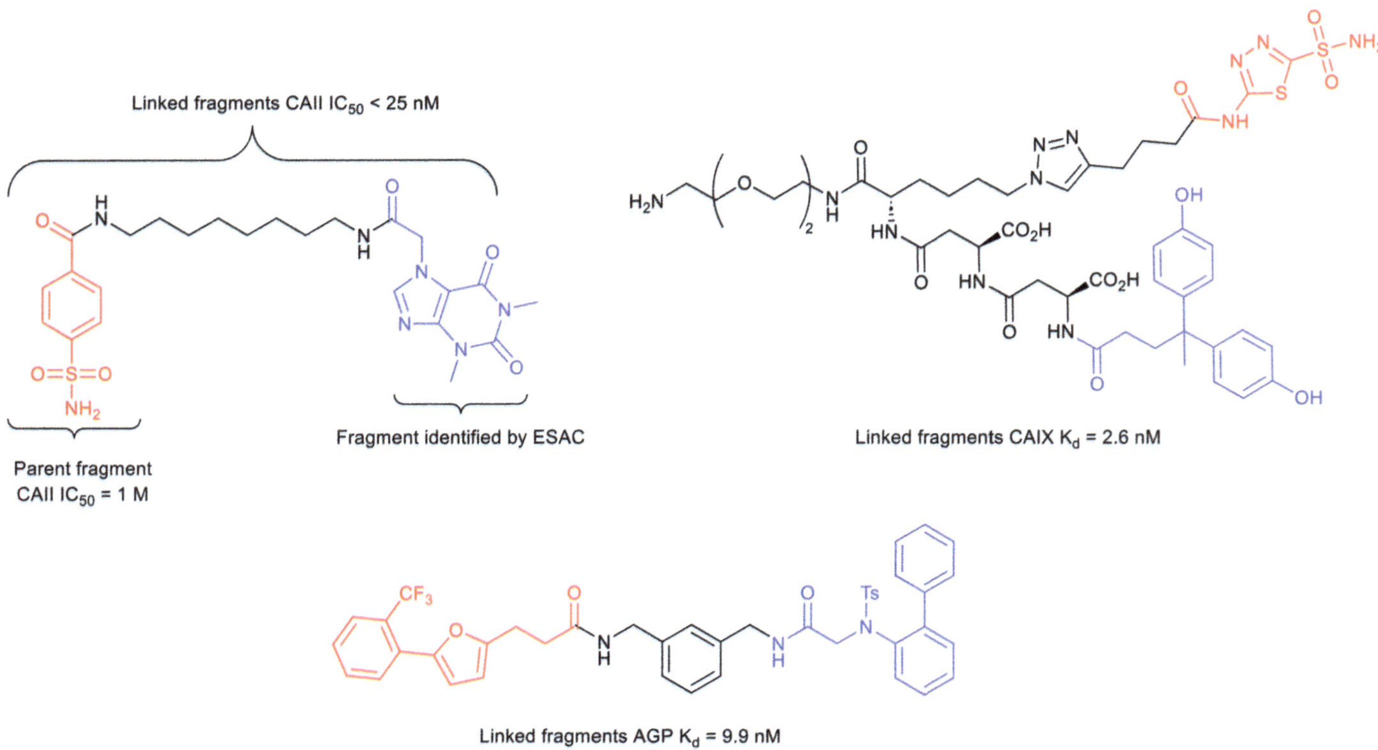

Figure 5.11 Exemplary structures of ligands identified *via* fragment expansion using ESAC libraries. Fragment pairs identified from library selection are highlighted in red (established fragment) or blue (enriched during selection); subsequent investigation to determine optimal linking of fragments was performed off-DNA.[49,52–56]

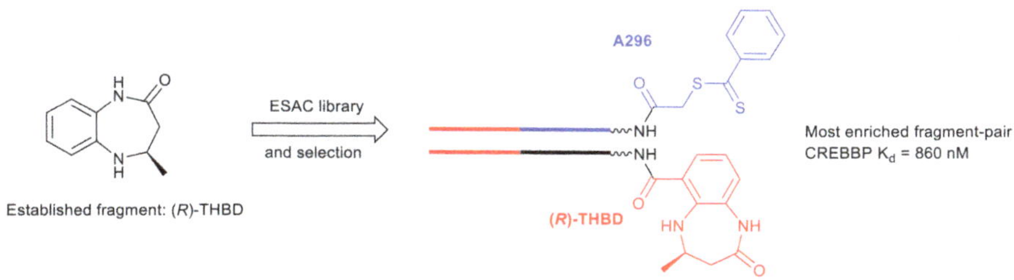

Figure 5.12 ESAC library employed for fragment expansion against CREBBP. Established fragment (*R*)-THBD was incorporated into a 335 000-member library and selection against the target identified the ligand pair (*R*)-THBD/A296 as the most enriched structure. K_d measurements of the fragment pairs were performed on modified LNA oligonucleotide derivatives *via* fluorescence polarization.[57] Adapted from ref. 57 with permission from John Wiley & Sons, Copyright 2020 Wiley-VCH GmbH.

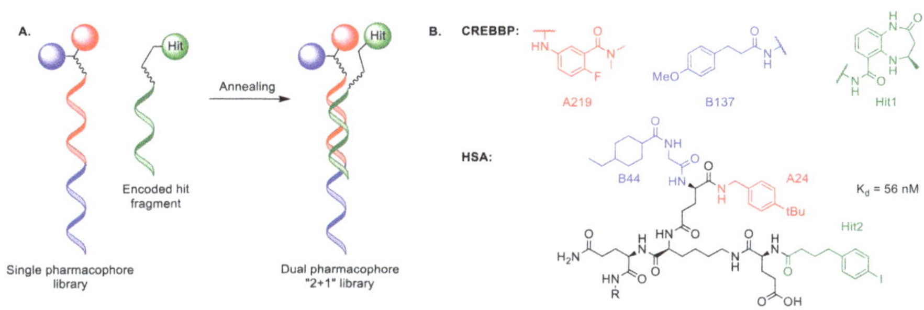

Figure 5.13 Overview of the "2+1" methodology proposed by Neri *et al.* for fragment expansion from DELs. (A) General architecture of the "2+1" library containing a single-stranded split-and-pool library annealed with a fragment hit. (B) Fragment pairs identified through the exemplification of the technique.[58] Adapted from ref. 58 https://doi.org/10.1002/advs.202001970, under the terms of the CC BY 4.0 license, https://creativecommons.org/licenses/by/4.0/.

combination of identified fragments displaying a 30-fold affinity enhancement in contrast to (*R*)-THBD in conjunction with an acetyl moiety (K_d = 860 nM). Further development of the fragment pair into a conjugated ligand has not been reported.

The same year, Neri *et al.* also proposed a method for the synthesis of a single-stranded split-and-pool library that had the potential to be hybridised with an established fragment on the complementary strand, thus enabling one DEL to be employed in both a single-pharmacophore *de novo* ligand identification and a dual-pharmacophore fragment expansion manner, a method the authors termed a "2+1" approach (Figure 5.13).[58] Exemplifying the library use as a fragment expansion tool, the group employed the previously discussed THBD/CREBBP system, demonstrating that the dual-pharmacophore fragment-expanded hit showed improved affinity over the

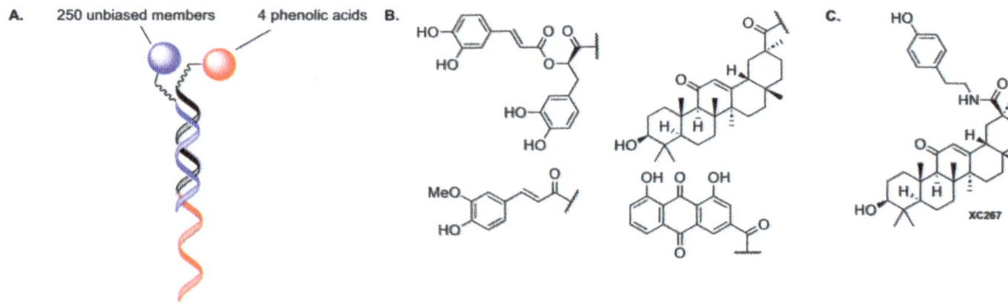

Figure 5.14 Overview of the ESAC methodology employed by Zhao *et al.* for expansion of natural product fragments for potential asthma treatments. (A) General architecture of the ESAC library. (B) Structures of established fragments employed. (C) Structure of XC267, a dual-target ligand identified from selections.[59,60] Adapted from ref. 59 with permission from Elsevier, Copyright 2022.

single-pharmacophore combination of building blocks. As a secondary validation, selection was also performed against human serum albumin (HSA) utilising 4-(*p*-iodophenyl)butanoic acid as the fragment of interest. Enrichment was observed for a pair of building blocks A24/B137 in combination with the fragment and subsequent linking identified a structure with a K_d of 56 nM. Whilst this method demonstrates potential for combinatorial expansion of fragments from ESAC libraries, ultimately the compounds generated are significantly outside of druglike chemical space, limiting the utility for hit identification.

Zhao *et al.* employed an ESAC-type library in their efforts to expand upon natural product fragments in drug discovery, in this instance for the development of potential asthma treatments (Figure 5.14).[59,60] Using four established fragments on one oligonucleotide strand (rosmarinic acid, 18β-glycyrrhetinic acid, rhein and ferulic acid), the second strand was conjugated with 250 varied acids, generating a 1000-member library. Selection was performed separately against two relevant asthma targets with a synergistic effect, β_2-adrenoreceptor (β_2-AR) and cysteinyl-leukotriene receptor (CysLT), with the aim of identifying compounds with the ability to bind to both. XC267 was identified as showing enrichment in both selections, and off-DNA validation of the compound showed that it possessed similar or improved binding affinities to several established treatments currently in use.

Significant recent advances have been made within the ESAC field for fragment expansion; whilst evidently conventional ESAC DEL selection has been beneficial for the identification of pairs of fragments, a significant amount of follow-up work is usually required to identify optimum linkages between the pair, impeding the rate of hit discovery and preventing the technique from having maximal impact. Accordingly, techniques that allow for the identification of optimised linkers during selection would result in significant benefit.

In 2023, Zhang *et al.* reported the use of a trio-pharmacophore DEL (T-DEL), in which the central pharmacophore is designed to function as a

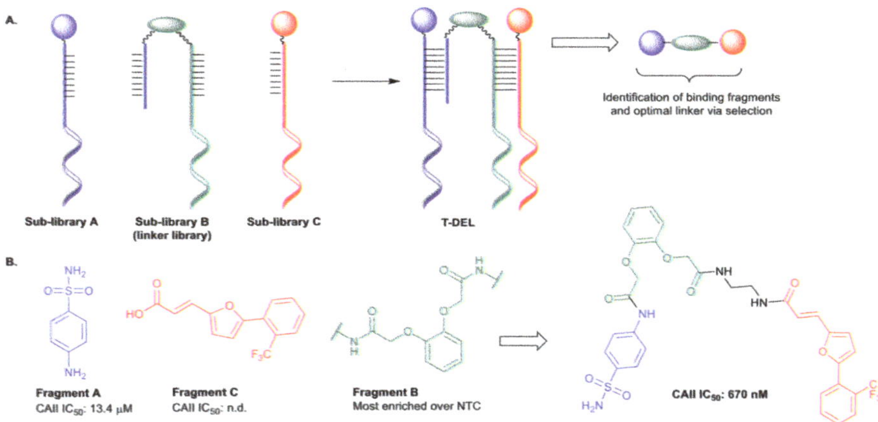

Figure 5.15 Overview of the T-DEL methodology developed by Zhang *et al.* for the simultaneous identification of binding fragments and optimal linker lengths from ESAC-type libraries. (A) General architecture of T-DEL containing three sub-libraries (SL) of fragments (SL-A, SL-C) and linkers (SL-B). Annealing of partially complementary strands between the sub-libraries allows for preferred fragments and linker length/geometry to be identified during selections. (B) Exemplification of the T-DEL approach employing an established binding pair for CAII in the presence of a 30-member SL-B.[61] Adapted from ref. 61 https://doi.org/10.1038/s41467-023-37071-1, under the terms of the CC BY 4.0 license, https://creativecommons.org/licenses/by/4.0/.

linker between the two fragment libraries (Figure 5.15).[61] The group demonstrated the method in selections against carbonic anhydrase II (CAII) and bovine trypsin, generating a 30-member proof-of-concept DEL in each instance with known fragment pairs of interest and varied encoded-linkers, before employing the method in the synthesis of a >23-million-member library. Whilst the technology has the potential to offer benefits for fragment expansion, its full impact is yet to be explored; other than the model T-DEL for CAII, no ligands exhibiting enrichment showed sub-μM activity for targets in off-DNA validation.

Li *et al.* proposed an alternative dynamic ESAC method in 2024 for determination of optimal linker length during the selection process in conjunction with fragments of interest.[62] In contrast to the previous method of incorporating separately encoded linker regions, the group hypothesised that enabling a cross-linking event between the fragments of interest would allow for the identification of the entire small molecule structure, including the linker (Figure 5.16). Using streptavidin as a proof-of-concept target, a model library was constructed, containing desthiobiotin in conjunction with various azide/alkyne linkers (dependent on the strand) to permit a click reaction to occur when oriented in optimal proximity. Results demonstrated that the cycloaddition between the two fragments was specific and occurred only when oriented correctly. A second proof-of-concept selection was

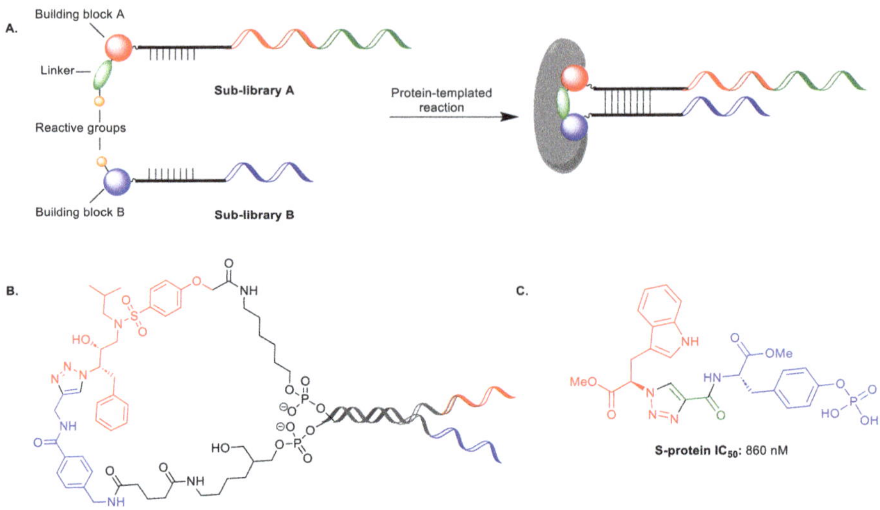

Figure 5.16 Overview of the protein templated cross-linking methodology described by Li *et al.* for identification of binding fragments and linkers from ESAC-type libraries. (A) Structure of library members including encoded fragments and linker of interest containing reactive groups (alkyne/azide) at appropriate positions to facilitate cross-linking. In the presence of the target, library members are subjected to $CuSO_4$, NaAs, and THPTA to permit a click reaction between fragments located at the optimal orientations. (B) Structure of the AB3 oligo-conjugate employed as validation of the methodology. The structure was originally present as the corresponding azide/alkyne fragments and cross-linking was only observed in the presence of the known target HIV-1 protease. (C) Structure of the SARS-CoV-2 spike glycoprotein ligand identified from selection of a full-scale library using the described technique.[62] Adapted from ref. 62 with permission from Springer Nature, Copyright 2024.

performed employing HIV-1 protease and AB3 as the ligand. Segmentation of AB3 into relevant alkyne/azide pairs was performed, and cross-linking was only observed in selections of HIV-1 protease, with control experiments (alternative proteins, beads only) exhibiting no fragment conjugation. The method was subsequently incorporated into two full-scale libraries, containing 4.35 million and 3 million compounds, respectively. Selections were performed across a variety of targets, with the most potent compound identified as a 860 nM inhibitor of SARS-CoV-2 spike glycoprotein (S-protein).

Whilst more prevalent within the ESAC field, conventional split-and-pool DELs have also been exploited for fragment expansion. Li *et al.* proposed a dynamic fragment-directed technique relying on *in situ* hit formation with an appropriately derivatised library in 2020 (Figure 5.17).[63] Employing an aldehyde derivative of the fragment of interest, the group generated several peptide-based libraries (440 to 17.5 million members) terminated with a primary amine, enabling reversible imine formation with the fragment.

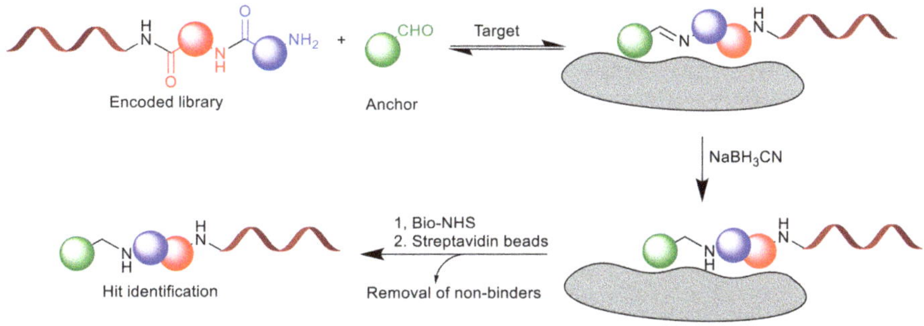

Figure 5.17 Dynamic fragment-directed methodology proposed by Li *et al.* for fragment expansion from DELs. Employing the aldehyde derivative of an established fragment hit ("anchor") in the presence of an amino-terminated DEL, the group developed a technique for *in situ* hit formation and identification.[63] Adapted from ref. 63 with permission from John Wiley & Sons, Copyright 2020 Wiley-VCH Verlag GmbH & Co. KGaA, Weinheim.

In the presence of the target protein, binding of the fragment will promote formation of the highest affinity compounds, and the equilibrium can be quenched through $NaBH_3CN$-mediated reduction. Residual primary amines are then captured *via* the addition of a biotin sulfo-*N*-hydroxy succinimidyl ester and subsequent streptavidin bead immobilisation.

Whilst the method offers the potential benefit of removing the requirement of generating bespoke-libraries for targets of interest, instead enabling DELs to be targeted *in situ*, it does rely on a number of caveats that the authors themselves recognise. Firstly, removal of inactive primary amines through reaction with derivatised biotin requires all amines to exhibit equivalent reactivity, potentially limiting the structural diversity of building blocks that may be employed to prevent the identification of false positives. Additionally, the method relies on the generated secondary amino hits being captured inefficiently enough that sufficient material remains for their detection. Whilst the authors did not report experiencing either of these issues during the model selections, it is unclear if this will always be the case. Finally, the binding compounds during the selection are imines, in contrast to the synthesised hit compounds which are amines; the methodology thus assumes that the amine counterparts will exhibit the same affinity for the target as their respective imines, which is unlikely in most cases.

Waring *et al.* adopted a related static approach for fragment expansion from DELs involving the preparation of poised libraries, members of which contained a reactive group which could be coupled with an active fragment hit of choice (Figure 5.18).[64] Subsequent selection of the libraries would thus allow for rapid optimisation of the fragment into a more potent, lead-like structure. To exemplify the approach, a 42-member proof-of-concept poised library was synthesised by sequential amide couplings, leaving a halide in place as the reactive handle. Subsequent Suzuki–Miyaura coupling to the

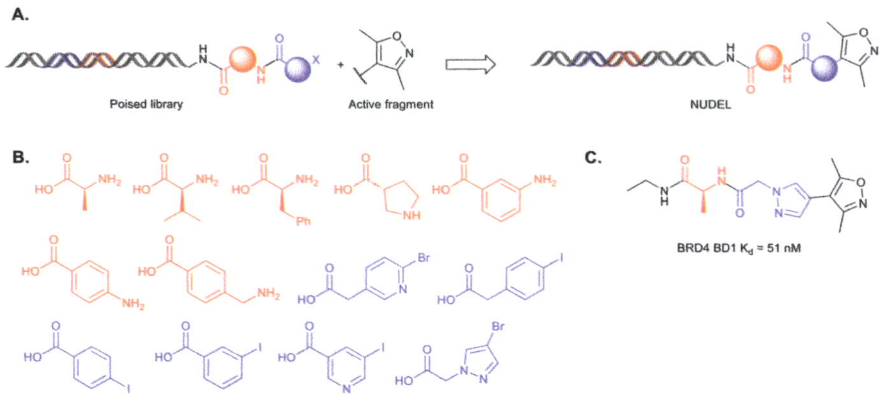

Figure 5.18 Overview of NUDEL technology proposed by Waring *et al.* for fragment expansion from DELs. (A) General architecture of libraries consisting of generation of a poised library terminated with a reactive handle which can be coupled to an active fragment for the target of interest. (B) Structures of building blocks employed in the synthesis of the prototypical library (NUDEL). (C) Structure of the ligand identified from selection of NUDEL against BRD4 BD1.[64] Adapted from ref. 64 with permission from the Royal Society of Chemistry.

corresponding boronic acid of 3,5-dimethylisoxazole, a KAc-mimetic and known active fragment for BRD4 completed the desired library. Selection against the target identified significant enrichment for a pair of building blocks, with neither showing enhancement in any other combination suggesting that the synergistic combination of the two motifs was responsible for the gain in affinity, a result that could not have been garnered *via* conventional fragment growing. Hit identity and affinity were confirmed off-DNA with SPR and X-ray crystallography, with the obtained ligand showing a K_d of 51 nM for BRD4, demonstrating the potential of this approach for rapid fragment expansion.

5.4 Hit Optimisation

DELs have also shown promise in the application to follow-up hit-to-lead strategies. A typical DEL workflow subsequent to selection is exemplified by Xiao *et al.* in their pursuit of selective PDE3B inhibitors. After screening of the GSK collection of $>10^{12}$ encoded compounds, a single series containing a boronic acid-based building block was identified as possessing the desired target selectivity. Subsequent exploration of the class was performed *via* conventional SAR off-DNA.[65] More recently, several groups have demonstrated that this optimisation, with respect to both target affinity and selectivity, can be achieved on-DNA. In this manner, an initial less-focused library is selected against a target to identify a ligand(s) of interest; these ligands are then employed as a template for construction of a secondary DEL, which is re-selected against the target to identify improved hits.

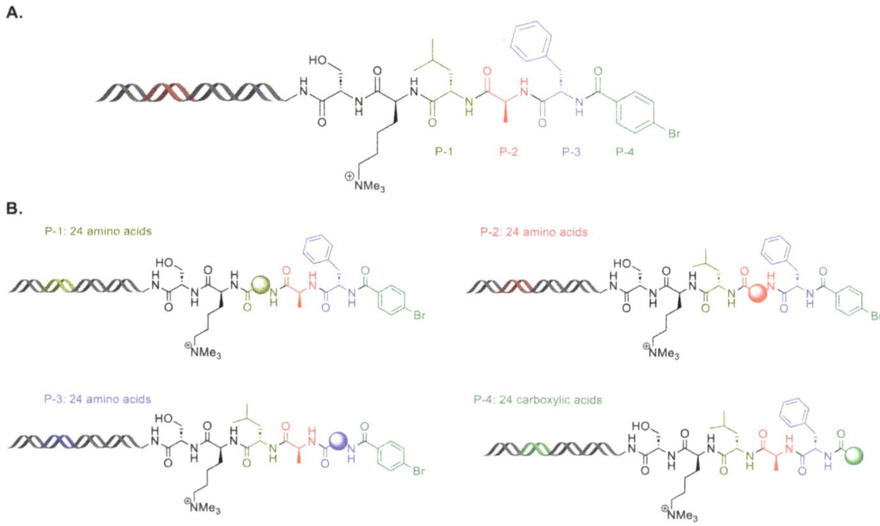

Figure 5.19 Positional scanning approach employed by Krusemark *et al.* for identification of selective CBX8 inhibitors. (A) Original hit structure on which subsequent libraries were based. (B) Architectures of the first positional scanning library indicating the point of diversity.[66] Adapted from ref. 66 with permission from Elsevier, Copyright 2018.

Krusemark *et al.* were the first group to successfully demonstrate that DELs can be employed for hit optimisation through their use of a positional scanning approach in their efforts to identify selective inhibitors for the highly conserved chromodomains of chromobox (CBX) proteins (Figure 5.19).[66] Employing an established hexapeptide ligand (CBX7; K_d = 97 nM) as the library template, each of the four terminal positions was varied in an independent manner to generate a 96-member library. Selection against CBX7 and CBX8 demonstrated that significant SAR could be garnered from this approach, with respect to both favourable potency and selectivity for each homolog.

This library was then employed as the starting point in a campaign to identify selective cell-permeable inhibitors against CBX8, expanding selections to include the chromodomains of CBX2, CBX4, CBX5 and CBX6 (Figure 5.20).[67] Results obtained from this initial selection echoed those observed previously, with modifications at P-2 being the most critical position for obtaining selectivity. Based on the most selective ligand obtained from this selection (KED98), a subsequent positional scanning library was synthesised in an effort to improve the affinity, selectivity and cellular permeability of this hit. In this library, alongside independent modification at each of the previously explored positions, several monomers were also incorporated that corresponded to two positions in the parental ligand (*i.e.* P1+2, P2+3, P3+4). Selection was performed under both normal and high stringency conditions, yielding further information regarding tolerance and selectivity of various modifications across

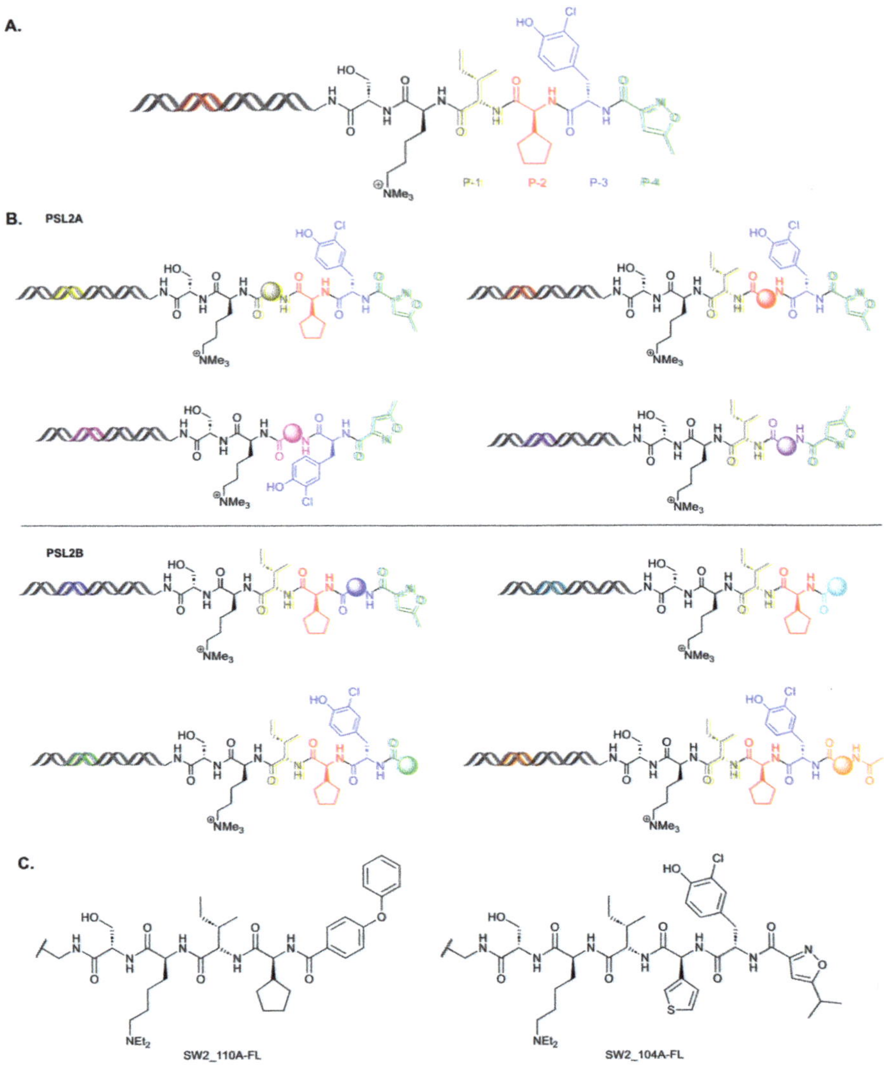

Figure 5.20 Second positional scanning library for identification of selective CBX8 inhibitors. (A) Hit structure from previous selections on which library synthesis was based. (B) General architectures of follow-up libraries built for hit optimisation. PSL-2A contained modifications at P1, P2, P1+2 and P2+3. PSL-2B contained modifications at P3, P4, P3+4 and P4+Ac. (C) Hits identified with improved selectivity against CBX8 and greater affinity. IC_{50} values measured as fluorescein conjugates.[67] Adapted from ref. 67 with permission from American Chemical Society, Copyright 2020.

the CBX family and ultimately leading to the identification of both a highly selective CBX8 inhibitor (SW2_110A-FL, K_d = 800 nM) and SW2_104A-FL, which showed improved affinity for CBX8 (K_d = 2.9 nM), but less selectivity over other paralogs.[67]

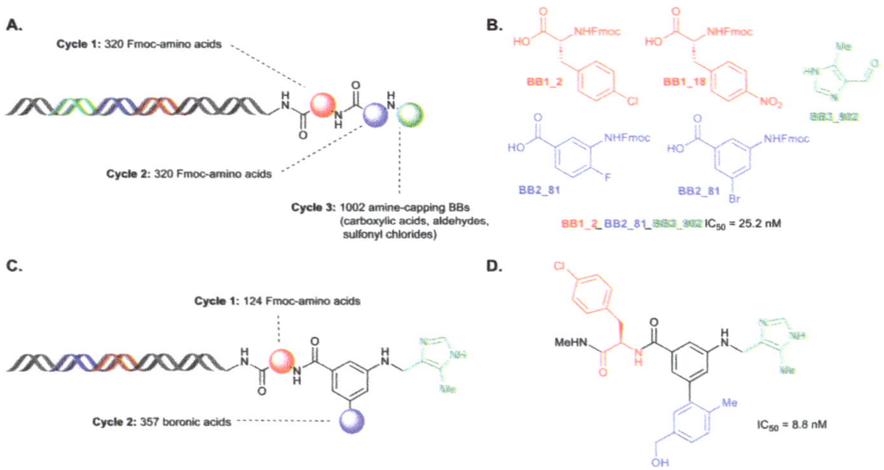

Figure 5.21 Overview of the cascade approach employed by Lu *et al.* for the discovery and optimisation of WDR5 inhibitors. (A) Initial library design employed for selections. (B) Building blocks of interest identified from primary screen and the IC_{50} value of the most potent combination. (C) Revised library design for hit optimisation through further selections. (D) Hit structure identified from cascade campaign.[68] Adapted from ref. 68 with permission from American Chemical Society, Copyright 2024.

More recently, Lu *et al.* adopted a cascade selection approach for the discovery and optimisation of inhibitors against WDR5 (Figure 5.21).[68] Initial selections of a 102.8-million-member unbiased linear library demonstrated preferential enrichment for 5-methyl-1*H*-imidazole as the terminal building block in conjunction with particular BB1/BB2 combinations. Representative hit compounds were validated off-DNA in an HTRF assay, with all demonstrating nM inhibition of the WDR5/MLL peptide complex (IC_{50} = 25–240 nM). Subsequent SAR study and molecular docking into the WIN site of WDR5 suggested that improved potency could be achieved through the design of branching structures, allowing growth into the S4 pocket. Accordingly, a second focused branched DEL was synthesised and selected against WDR5 in order to rapidly explore such structures. Results of this subsequent campaign identified multiple compounds with improved *in vitro* activity, with the most potent demonstrating a 3-fold improvement over the original hit (IC_{50} = 8.8 nM), exemplifying the potential of a cascade-based approach for hit optimisation.[68]

5.5 Future Directions

The applications of DELs have certainly broadened since the first reported libraries were synthesised, with significant expansion in the past five-years. No longer restricted solely to a method for high-throughput screening for initial hit identification, as evidenced in this chapter, the role of DELs has evolved for a variety of purposes within drug discovery. However, there are still a number of limitations to medicinal chemistry that can be performed on-DNA.

The emergence of focused or targeted encoded libraries has in some instances demonstrated that the method possesses some benefits over conventional unbiased DEL screening. Imbuing a library with attributes designed to enhance interactions with a specific target should certainly improve the hit identification rate and lead to the discovery of new ligands; the protease-focused library reported by Matzuk *et al.* is indeed an excellent example of this.[12] However, the successful application of a targeted library is highly reliant on the chemistry required for its synthesis. Many of the focused libraries reported thus far are constructed *via* sequential amide bond formation, which may allow for the identification of potent ligands; however, the inclusion of multiple amide bonds could, in some cases, result in compounds with compromised properties such as poor permeability or plasma instability requiring significant optimisation for development into therapeutic candidates. The availability of more high yielding universally applicable DNA-compatible chemistry would help to address this issue, leading to libraries containing more drug-like structures, which could be appropriately focused and require less optimisation post-hit identification.

Furthermore, in terms of the incorporation of a recognition element into a library, with the commercially available building blocks amenable to DEL synthesis, this can only be achieved readily in the terminal coupling step. Excluding amino acids, there is currently a limited number of suitably derivatised bifunctional building blocks available for DEL construction, and thus branched targeted libraries are restricted in terms of both size and diversity that can be incorporated. As an additional point, whilst there are obvious benefits to focusing DELs *via* this approach, appreciation of the target under consideration is a necessity. Whilst several of the focused DELs discussed here have been employed for the identification of agents to help combat antimicrobial resistance, it could be argued that novel modalities with unique mechanisms of action would be of greater benefit in this area, rather than libraries containing structures that are present in established therapies. Palzkill *et al.* demonstrated this approach in a beneficial way when using alternative pharmacophores to those present in conventional antibiotics.

Whilst methodology towards scaffold-focused libraries has been described, many of the reactions developed suffer from a limited substrate scope and/or significant building block variation.[24–39] Such features significantly impact either the diversity of the DEL if such building blocks are to be excluded during library synthesis, or the reliability of the DEL if they are incorporated. Both of these outcomes will negatively affect the hit identification rate from such libraries, reiterating the importance of development of DNA-compatible chemistry.

The ability to employ DELs for the identification of encoded fragments using the method devised by Liu *et al.* offers significant promise given the potential size of fragment libraries that could be generated and the corresponding chemical space explored. This does, however, still show reliance on both building block availability and suitable chemistry for attachment to oligonucleotides. One could imagine a cascade approach in which an encoded fragment library is employed to first identify a fragment of interest, which is then incorporated into a targeted/fragment expansion DEL, streamlining a significant portion of the hit-to-lead process.

With respect to fragment expansion, whilst a number of novel methods are emerging to allow for this technique, many are still relatively early on in development with overreliance on multiple caveats that limit their impact or the incorporation of compounds that sufficiently deviate from useful drug-like space. It could however be envisioned that methods related to the combined identification of fragments and linkers in ESAC could be applied to the discovery of CIP-type compounds. Depending on the strategy employed in this instance, it is likely that more versatile chemistry for linker installation would be required. Furthermore, Zhao *et al.* reported a method enabling conversion between double- and single-stranded DEL formats; as split-and-pool double-stranded DELs are the most common within the field, such a technique could attract application of these products in an ESAC "2+1" approach in conjunction with a fragment, although the utility of the structures identified in these instances is unclear.[69]

The potential for hit optimisation using DELs and cascade library approaches is somewhat promising; however, due to the nature of selection methods, currently the only compound metric that could be optimised in this manner is affinity. Whilst evidently an important property of any potential therapeutic candidate, such a feature is not the sole defining characteristic requiring optimisation and as such, until alternative selection methods that permit optimisation of other features are developed, such an avenue is restricted. Ekins *et al.* have suggested the combination of DEL selections with machine learning (ML) could be employed for expansion of the types of optimisations that could be achieved, giving examples such as toxicology predictions and cell-permeability.[70]

Ultimately however, one of the biggest challenges for performing medicinal chemistry on-DNA is the ability to correlate compound affinity with enrichment. Whilst the ability to measure the affinity of isolated DNA-hit conjugates and correlation with their off-DNA counterparts has been demonstrated in multiple instances, such an endeavour is significantly complicated in library selections.[12,71] Disproportionate initial member populations due to varied recoveries and/or incomplete synthetic transformations during library construction are one example of bias that can be introduced that clouds deconvolution of selection results. There are numerous reports claiming to address this issue computationally; however, at present a general consensus does not appear to have been reached.[23,72,73]

References

1. S. Brenner and R. A. Lerner, Encoded combinatorial chemistry, *Proc. Natl. Acad. Sci. U. S. A.*, 1992, **89**, 5381–5383.
2. P. S. Dragovich, W. Haap, M. M. Mulvihill, J.-M. Plancher and A. F. Stepan, Small-Molecule Lead-Finding Trends across the Roche and Genentech Research Organizations, *J. Med. Chem.*, 2022, **65**, 3606–3615.
3. M. A. Clark, R. A. Acharya, C. C. Arico-Muendel, S. L. Belyanskaya, D. R. Benjamin, N. R. Carlson, P. A. Centrella, C. H. Chiu, S. P. Creaser, J. W. Cuozzo, C. P. Davie, Y. Ding, G. J. Franklin, K. D. Franzen, M. L. Gefter, S. P. Hale, N. J. V. Hansen, D. I. Israel, J. Jiang, M. J. Kavarana, M. S. Kelley, C. S. Kollmann, F. Li, K. Lind, S. Mataruse, P. F. Medeiros, J. A. Messer, P. Myers, H. O'Keefe, M. C. Oliff, C. E. Rise, A. L. Satz, S. R. Skinner, J. L. Svendsen, L. Tang, K. van Vloten, R. W. Wagner, G. Yao, B. Zhao and B. A. Morgan, Design, synthesis and selection of DNA-encoded small-molecule libraries, *Nat. Chem. Biol.*, 2009, **5**, 647–654.
4. P. A. Harris, B. W. King, D. Bandyopadhyay, S. B. Berger, N. Campobasso, C. A. Capriotti, J. A. Cox, L. Dare, X. Dong, J. N. Finger, L. C. Grady, S. J. Hoffman, J. U. Jeong, J. Kang, V. Kasparcova, A. S. Lakdawala, R. Lehr, D. E. McNulty, R. Nagilla, M. T. Ouellette, C. S. Pao, A. R. Rendina, M. C. Schaeffer, J. D. Summerfield, B. A. Swift, R. D. Totoritis, P. Ward, A. Zhang, D. Zhang, R. W. Marquis, J. Bertin and P. J. Gough, DNA-Encoded Library Screening Identifies Benzo[*b*][1,4]oxazepin-4-ones as Highly Potent and Monoselective Receptor Interacting Protein 1 Kinase Inhibitors, *J. Med. Chem.*, 2016, **59**, 2163–2178.
5. P. A. Harris, S. B. Berger, J. U. Jeong, R. Nagilla, D. Bandyopadhyay, N. Campobasso, C. A. Capriotti, J. A. Cox, L. Dare, X. Dong, P. M. Eidam, J. N. Finger, S. J. Hoffman, J. Kang, V. Kasparcova, B. W. King, R. Lehr, Y. Lan, L. K. Leister, J. D. Lich, T. T. MacDonald, N. A. Miller, M. T. Ouellette, C. S. Pao, A. Rahman, M. A. Reilly, A. R. Rendina, E. J. Rivera, M. C. Schaeffer, C. A. Sehon, R. R. Singhaus, H. H. Sun, B. A. Swift, R. D. Totoritis, A. Vossenkämper, P. Ward, D. D. Wisnoski, D. Zhang, R. W. Marquis, P. J. Gough and J. Bertin, Discovery of a First-in-Class Receptor Interacting Protein 1 (RIP1) Kinase Specific Clinical Candidate (GSK2982772) for the Treatment of Inflammatory Diseases, *J. Med. Chem.*, 2017, **60**, 1247–1261.
6. S. L. Belyanskaya, Y. Ding, J. F. Callahan, A. L. Lazaar and D. I. Israel, Discovering Drugs with DNA-Encoded Library Technology: From Concept to Clinic with an Inhibitor of Soluble Epoxide Hydrolase, *ChemBioChem*, 2017, **18**, 837–842.
7. J. W. Cuozzo, M. A. Clark, A. D. Keefe, A. Kohlmann, M. Mulvihill, H. Ni, L. M. Renzetti, D. I. Resnicow, F. Ruebsam, E. A. Sigel, H. A. Thomson, C. Wang, Z. Xie and Y. Zhang, Novel Autotaxin Inhibitor for the Treatment of Idiopathic Pulmonary Fibrosis: A Clinical Candidate Discovered Using DNA-Encoded Chemistry, *J. Med. Chem.*, 2020, **63**, 7840–7856.

8. M. Korn, C. Ehrt, F. Ruggiu, M. Gastreich and M. Rarey, Navigating large chemical spaces in early-phase drug discovery, *Curr. Opin. Struct. Biol.*, 2023, **80**, 102578.
9. Y. Zhang and R. M. Franzini, Design Considerations in Constructing and Screening DNA-Encoded Libraries, *Top. Med. Chem.*, 2022, **40**, 123–143.
10. Y. Zhang and M. A. Clark, Design concepts for DNA-encoded library synthesis, *Bioorg. Med. Chem.*, 2021, **41**, 116189.
11. L. H. Yuen, S. Dana, Y. Liu, S. I. Bloom, A.-G. Thorsell, D. Neri, A. J. Donato, D. Kireev, H. Schüler and R. M. Franzini, A Focused DNA-Encoded Chemical Library for the Discovery of Inhibitors of NAD^+-Dependent Enzymes, *J. Am. Chem. Soc.*, 2019, **141**, 5169–5181.
12. S. Dawadi, N. Simmons, G. Miklossy, K. M. Bohren, J. C. Faver, M. N. Ucisik, P. Nyshadham, Z. Yu and M. M. Matzuk, Discovery of potent thrombin inhibitors from a protease-focused DNA-encoded chemical library, *Proc. Natl. Acad. Sci. U. S. A.*, 2020, **117**, 16782–16789.
13. D. J. Shell, C. A. Foley, Q. Wang, C. M. Smith, S. K. R. Guduru, H. Zeng, A. Dong, J. L. Norris-Drouin, M. Axtman, P. B. Hardy, G. Gupta, L. Halabelian, S. V. Frye, L. I. James and K. H. Pearce, Discovery of a 53BP1 Small Molecule Antagonist Using a Focused DNA-Encoded Library Screen, *J. Med. Chem.*, 2023, **66**, 14133–14149.
14. J. M. Rectenwald, S. K. Guduru, Z. Dang, L. B. Collins, Y.-E. Liao, J. L. Norris-Drouin, S. H. Cholensky, K. W. Kaufmann, S. M. Hammond, D. B. Kireev, S. V. Frye and K. H. Pearce, *Molecules*, 2020, **25**, 979.
15. D. J. Shell, J. M. Rectenwald, P. H. Buttery, R. L. Johnson, C. A. Foley, S. K. R. Guduru, M. Uguen, J. S. Rubiano, X. Zhang, F. Li, J. L. Norris-Drouin, M. Axtman, P. Brian Hardy, M. Vedadi, S. V. Frye, L. I. James and K. H. Pearce, Discovery of hit compounds for methyl-lysine reader proteins from a target class DNA-encoded library, *SLAS Discovery*, 2022, **27**, 428–439.
16. Q. Liang, J. He, X. Zhao, Y. Xue, H. Zuo, R. Xu, Y. Jin, J. Wang, Q. Li and X. Zhao, Selective Discovery of GPCR Ligands within DNA-Encoded Chemical Libraries Derived from Natural Products: A Case Study on Antagonists of Angiotensin II Type I Receptor, *J. Med. Chem.*, 2021, **64**, 4196–4205.
17. Q. M. R. Gibaut, Y. Akahori, J. A. Bush, A. Taghavi, T. Tanaka, H. Aikawa, L. S. Ryan, B. M. Paegel and M. D. Disney, Study of an RNA-Focused DNA-Encoded Library Informs Design of a Degrader of a r(CUG) Repeat Expansion, *J. Am. Chem. Soc.*, 2022, **144**, 21972–21979.
18. W. G. Cochrane, P. R. Fitzgerald and B. M. Paegel, Antibacterial Discovery *via* Phenotypic DNA-Encoded Library Screening, *ACS Chem. Biol.*, 2021, **16**, 2752–2756.
19. P. R. Fitzgerald, W. G. Cochrane and B. M. Paegel, Dose-Response Activity-Based DNA-Encoded Library Screening, *ACS Med. Chem. Lett.*, 2023, **14**, 1295–1303.
20. M. Békés, D. R. Langley and C. M. Crews, PROTAC targeted protein degraders: the past is prologue, *Nat. Rev. Drug Discovery*, 2022, **21**, 181–200.

21. J. W. Mason, Y. T. Chow, L. Hudson, A. Tutter, G. Michaud, M. V. Westphal, W. Shu, X. Ma, Z. Y. Tan, C. W. Coley, P. A. Clemons, S. Bonazzi, F. Berst, K. Briner, S. Liu, F. J. Zecri and S. L. Schreiber, DNA-encoded library-enabled discovery of proximity-inducing small molecules, *Nat. Chem. Biol.*, 2024, **20**, 170–179.
22. S. Park, J. Fan, S. Chamakuri, M. Palaniappan, K. Sharma, X. Qin, J. Wang, Z. Tan, A. Judge, L. Hu, B. Sankaran, F. Li, B. V. V. Prasad, M. M. Matzuk and T. Palzkill, Exploiting the Carboxylate-Binding Pocket of β-Lactamase Enzymes Using a Focused DNA-Encoded Chemical Library, *J. Med. Chem.*, 2024, **67**, 620–642.
23. A. L. Satz, R. Hochstrasser and A. C. Petersen, Analysis of Current DNA Encoded Library Screening Data Indicates Higher False Negative Rates for Numerically Larger Libraries, *ACS Comb. Sci.*, 2017, **19**, 234–238.
24. S. Chamakuri, M.-K. Chung, E. L. G. Samuel, K. A. Tran, Y.-C. Chen, P. Nyshadham, C. Santini, M. M. Matzuk and D. W. Young, Design and construction of a stereochemically diverse piperazine-based DNA-encoded chemical library, *Bioorg. Med. Chem.*, 2021, **48**, 116387.
25. Q. Nie, X. Fang, C. Liu, G. Zhang, X. Fan, Y. Li and Y. Li, DNA-compatible ortho-phthalaldehyde (OPA)-mediated 2-substituted isoindole core formation and applications, *J. Org. Chem.*, 2022, **87**, 2551–2558.
26. J. Yuen, J. Chai and Y. Ding, Condensation of DNA-Conjugated Imines with Homophthalic Anhydride for the Synthesis of Isoquinolones On-DNA, *Methods Mol. Biol.*, 2022, **2541**, 75–79.
27. V. B. K. Kunig, M. Potowski, M. Akbarzadeh, M. Klika Skopic, D. dos Santos Smith, L. Arendt, I. Dormuth, H. Adihou, B. Andlovic, H. Karatas, S. Shaabani, T. Zarganes-Tzitzikas, C. G. Neochoritis, R. Zhang, M. Groves, S. M. Gueret, C. Ottmann, J. Rahnenfuehrer, R. Fried, A. Doemling and A. Brunschweiger, TEAD-YAP Interaction Inhibitors and MDM2 Binders from DNA-Encoded Indole-Focused Ugi Peptidomimetics, *Angew. Chem., Int. Ed.*, 2020, **59**, 20338–20342.
28. M. A. Mantell, L. Marcaurelle and Y. Ding, One Reaction Served Three Ways: The On-DNA Ugi 4C-3C Reaction for the Formation of Lactams, *Org. Lett.*, 2023, **25**, 1241–1245.
29. Y. Gao, Y. Sun, G. Zhao, G. Zhang, Y. Li and Y. Li, On-DNA Synthesis of Functionalized 4H-Pyran Scaffolds for Focused DNA-Encoded Chemical Libraries, *Org. Lett.*, 2022, **24**, 6664–6669.
30. X. Wen, M. Zhang, Z. Duan, Y. Suo, W. Lu, R. Jin, B. Mu, K. Li, X. Zhang, L. Meng, Y. Hong, X. Wang, H. Hu, J. Zhu, W. Song, A. Shen and X. Lu, Discovery, SAR Study of GST Inhibitors from a Novel Quinazolin-4(1*H*)-one Focused DNA-Encoded Library, *J. Med. Chem.*, 2023, **66**, 11118–11132.
31. B. Matsuo, S. Kim, S. T. Shreiber, G. Levitre, L. Li, E. A. Crane, E. J. McClain, E. A. Voight and G. A. Molander, Diversifying chemical space of DNA-encoded libraries: synthesis of 2-oxa-1-azaspiro(bicyclo[3.2.0])-heptanes on-DNA *via* visible light-mediated energy transfer catalysis, *Chem. Commun.*, 2023, **59**, 10964–10967.

32. Y. Ye, M. Berry, W. J. Bock, K. Cheng, S. K. Nair, C. S. Park, R. L. Patman, S. Sakata, M. Tran-Dube, J. S. Donaldson, G. Yang and G. Liu, Construction of Isoquinolone Scaffolds on DNA *via* Rhodium(III)-Catalyzed C–H Activation, *Org. Lett.*, 2024, **26**, 3338–3342.
33. A. Gy. Nemeth, L. Kollar, K. Nemeth, G. Schlosser, A. Minus and G. M. Keseru, On-DNA Synthesis of Multisubstituted Indoles, *Org. Lett.*, 2024, **26**, 2517–2522.
34. J. Zhang, X. Li, H. Wei, Y. Li, G. Zhang and Y. Li, Sequential DNA-Encoded Building Block Fusion for the Construction of Polysubstituted Pyrazoline Core Libraries, *Org. Lett.*, 2021, **23**, 8429–8433.
35. S. Eom, T. Kwon, D. Y. Lee, C. H. Park and H. J. Kim, Copper-Mediated Three-Component Reaction for the Synthesis of N-Acylsulfonamide on DNA, *Org. Lett.*, 2022, **24**, 4881–4885.
36. J. Zhang, J. Liu, G. Zhang, X. He, F. Xiong, X. Fan, Y. Li and Y. Li, Synthesis of Diacylhydrazine Derivatives Based on Tetrazole-Focused DNA-Encoded Library, *Org. Lett.*, 2024, **26**, 1094–1099.
37. X. Fang, T. Zhang, W. Fang, G. Zhang, Y. Li and Y. Li, Synthesis of Functionalized Triazoles on DNA *via* Azide-Acetonitrile 'Click' Reaction, *Org. Lett.*, 2023, **25**, 8326–8331.
38. A. Luo, N. Duchemin, X. Wang, H. Zhou, F. Zeng, X. Zhao, W. Yu, K. Yang and Y. Jin Hu, Development of On-DNA Thiophene Synthesis for DNA-encoded chemical libraries (DELs) Construction, *Chem. – Asian J.*, 2023, **18**, e202300458.
39. Y. Suo, M. Xu, M. Sun, W. Lu, X. Wang and X. Lu, Ruthenium-mediated [2 + 2 + 2] cyclization: A route to forge indane and isoindoline core and its application in DNA-encoded library technology, *Org. Lett.*, 2022, **24**, 9092–9096.
40. M. Potowski, V. B. K. Kunig, F. Losch and A. Brunschweiger, Synthesis of DNA-coupled isoquinolones and pyrrolidines by solid phase ytterbium- and silver-mediated imine chemistry, *Med. Chem. Commun.*, 2019, **10**, 1082–1093.
41. M. K. Škopić, H. Salamon, O. Bugain, K. Jung, A. Gohla, L. J. Doetsch, D. dos Santos, A. Bhat, B. Wagner and A. Brunschweiger, Acid- and Au(I)-mediated synthesis of hexathymidine-DNA-heterocycle chimeras, an efficient entry to DNA-encoded libraries inspired by drug structures, *Chem. Sci.*, 2017, **8**, 3356–3361.
42. S. Liu, J. Qi, W. Lu, X. Wang and X. Lu, Synthetic Studies toward DNA-Encoded Heterocycles Based on the On-DNA Formation of α,β-Unsaturated Ketones, *Org. Lett.*, 2021, **23**, 908–913.
43. V. B. K. Kunig, C. Ehrt, A. Dömling and A. Brunschweiger, Isocyanide Multicomponent Reactions on Solid-Phase-Coupled DNA Oligonucleotides for Encoded Library Synthesis, *Org. Lett.*, 2019, **21**, 7238–7243.
44. L. Mannocci, S. Melkko, F. Buller, I. Molnàr, J.-P. Gapian Bianké, C. E. Dumelin, J. Scheuermann and D. Neri, Isolation of Potent and Specific Trypsin Inhibitors from a DNA-Encoded Chemical Library, *Bioconjugate Chem.*, 2010, **21**, 1836–1841.

45. Q. Nie, J. Sun, X. Fang, X. He, F. Xiong, G. Zhang, Y. Li and Y. Li, Antimony salt-promoted cyclization facilitating on-DNA syntheses of dihydroquinazolinone derivatives and its applications, *Chin. Chem. Lett.*, 2023, **34**, 108132.
46. X. Fang, H. Liao, X. Fan, Y. Wang, H. Wang, G. Zhang, W. Fang, Y. Li and Y. Li, Incorporation of viridicatin alkaloid-like scaffolds into DNA-encoded chemical libraries, *Org. Biomol. Chem.*, 2023, **21**, 2162–2166.
47. K. E. Denton and C. J. Krusemark, Crosslinking of DNA-linked ligands to target proteins for enrichment from DNA-encoded libraries, *MedChemComm*, 2016, **7**, 2020–2027.
48. H. Ma, J. B. Murray, H. Luo, X. Cheng, Q. Chen, C. Song, C. Duan, P. Tan, L. Zhang, J. Liu, B. A. Morgan, J. Li, J. Wan, L. M. Baker, W. Finnie, L. Guetzoyan, R. Harris, N. Hendrickson, N. Matassova, H. Simmonite, J. Smith, R. E. Hubbard and G. Liu, PAC-FragmentDEL – photoactivated covalent capture of DNA-encoded fragments for hit discovery, *RSC Med. Chem.*, 2022, **13**, 1341–1349.
49. S. Melkko, J. Scheuermann, C. E. Dumelin and D. Neri, Encoded self-assembling chemical libraries, *Nat. Biotechnol.*, 2004, **22**, 568–574.
50. Y. Zhou, J. Peng, W. Shen and X. Li, Psoralen as an interstrand DNA crosslinker in the selection of DNA-Encoded dynamic chemical library, *Biochem. Biophys. Res. Commun.*, 2020, **533**, 215–222.
51. S. Oehler, L. Plais, G. Bassi, D. Neri and J. Scheuermann, Modular assembly and encoding strategies for dual-display DNA-encoded chemical libraries, *Chem. Commun.*, 2021, **57**, 12289–12292.
52. M. Wichert, N. Krall, W. Decurtins, R. M. Franzini, F. Pretto, P. Schneider, D. Neri and J. Scheuermann, Dual-display of small molecules enables the discovery of ligand pairs and facilitates affinity maturation, *Nat. Chem.*, 2015, **7**, 241–249.
53. J. Scheuermann, C. E. Dumelin, S. Melkko, Y. Zhang, L. Mannocci, M. Jaggi, J. Sobek and D. Neri, DNA-Encoded Chemical Libraries for the Discovery of MMP-3 Inhibitors, *Bioconjugate Chem.*, 2008, **19**, 778–785.
54. F. V. Reddavide, W. Lin, S. Lehnert and Y. Zhang, DNA-Encoded Dynamic Combinatorial Chemical Libraries, *Angew. Chem., Int. Ed.*, 2015, **54**, 7924–7928.
55. F. V. Reddavide, M. Cui, W. Lin, N. Fu, S. Heiden, H. Andrade, M. Thompson and Y. Zhang, Second generation DNA-encoded dynamic combinatorial chemical libraries, *Chem. Commun.*, 2019, **55**, 3753–3756.
56. M. Bigatti, A. Dal Corso, S. Vanetti, S. Cazzamalli, U. Rieder, J. Scheuermann, D. Neri and F. Sladojevich, Impact of a Central Scaffold on the Binding Affinity of Fragment Pairs Isolated from DNA-Encoded Self-Assembling Chemical Libraries, *ChemMedChem*, 2017, **12**, 1748–1752.
57. M. Catalano, M. Moroglu, P. Balbi, F. Mazzieri, J. Clayton, K. H. Andrews, M. Bigatti, J. Scheuermann, S. J. Conway and D. Neri, Selective Fragments for the CREBBP Bromodomain Identified from an Encoded Self-assembly Chemical Library, *ChemMedChem*, 2020, **15**, 1752–1756.

58. G. Bassi, N. Favalli, M. Vuk, M. Catalano, A. Martinelli, A. Trenner, A. Porro, S. Yang, C. L. Tham, M. Moroglu, W. W. Yue, S. J. Conway, P. K. Vogt, A. A. Sartori, J. Scheuermann and D. Neri, A Single-Stranded DNA-Encoded Chemical Library Based on a Stereoisomeric Scaffold Enables Ligand Discovery by Modular Assembly of Building Blocks, *Adv. Sci.*, 2020, **7**, 2001970.
59. Q. Liang, H. Zuo, T. Yang, J. Yin, X. Huang, J. Wang, Q. Li, G. Yin and X. Zhao, Discovery of dual-target ligands binding to beta2-adrenoceptor and cysteinyl-leukotriene receptor for the potential treatment of asthma from natural products derived DNA-encoded library, *Eur. J. Med. Chem.*, 2022, **233**, 114212.
60. R. Tian, J. Yin, Q. Yao, T. Wang, J. Chen, Q. Liang, Q. Li and X. Zhao, Development of an Allostery Responsive Chromatographic Method for Screening Potential Allosteric Modulator of Beta2-adrenoceptor from a Natural Product-Derived DNA-Encoded Chemical Library, *Anal. Chem.*, 2022, **94**, 9048–9057.
61. M. Cui, D. Nguyen, M. P. Gaillez, S. Heiden, W. Lin, M. Thompson, F. V. Reddavide, Q. Chen and Y. Zhang, Trio-pharmacophore DNA-encoded chemical library for simultaneous selection of fragments and linkers, *Nat. Commun.*, 2023, **14**, 1481.
62. Y. Zhou, W. Shen, Y. Gao, J. Peng, Q. Li, X. Wei, S. Liu, F. S. Lam, J. Mayol-Llinàs, G. Zhao, G. Li, Y. Li, H. Sun, Y. Cao and X. Li, Protein-templated ligand discovery *via* the selection of DNA-encoded dynamic libraries, *Nat. Chem.*, 2024, **16**, 543–555.
63. Y. Deng, J. Peng, F. Xiong, Y. Song, Y. Zhou, J. Zhang, F. S. Lam, C. Xie, W. Shen, Y. Huang, L. Meng and X. Li, Selection of DNA-Encoded Dynamic Chemical Libraries for Direct Inhibitor Discovery, *Angew. Chem., Int. Ed.*, 2020, **59**, 14965–14972.
64. C. L. A. Salvini, B. Darlot, J. Davison, M. P. Martin, S. J. Tudhope, S. Turberville, A. Kawamura, M. E. M. Noble, S. R. Wedge, J. J. Crawford and M. J. Waring, Fragment expansion with NUDELs – poised DNA-encoded libraries, *Chem. Sci.*, 2023, **14**, 8288–8294.
65. A. M. Rowley, G. Yao, L. Andrews, A. Bedermann, R. Biddulph, R. Bingham, J. J. Brady, R. Buxton, T. Cecconie, R. Cooper, A. Csakai, E. N. Gao, M. C. Grenier-Davies, M. Lawler, Y. Lian, J. Macina, C. Macphee, L. Marcaurelle, J. Martin, P. McCormick, R. Pindoria, M. Rauch, W. Rocque, Y. Shen, L. M. Shewchuk, M. Squire, W. Stebbeds, W. Tear, X. Wang, P. Ward and S. Xiao, Discovery and SAR Study of Boronic Acid-Based Selective PDE3B Inhibitors from a Novel DNA-Encoded Library, *J. Med. Chem.*, 2024, **67**, 2049–2065.
66. K. E. Denton, S. Wang, M. C. Gignac, N. Milosevich, F. Hof, E. C. Dykhuizen and C. J. Krusemark, Robustness of In Vitro Selection Assays of DNA-Encoded Peptidomimetic Ligands to CBX7 and CBX8, *SLAS Discovery*, 2018, **23**, 417–428.
67. S. Wang, K. E. Denton, K. F. Hobbs, T. Weaver, J. M. B. McFarlane, K. E. Connelly, M. C. Gignac, N. Milosevich, F. Hof, I. Paci, C. A. Musselman,

E. C. Dykhuizen and C. J. Krusemark, Optimization of Ligands Using Focused DNA-Encoded Libraries To Develop a Selective, Cell-Permeable CBX8 Chromodomain Inhibitor, *ACS Chem. Biol.*, 2020, **15**, 112–131.
68. S. Qin, L. Feng, Q. Zhao, Z. Yan, X. Lyu, K. Li, B. Mu, Y. Chen, W. Lu, C. Wang, Y. Suo, J. Yue, M. Cui, Y. Li, Y. Zhao, Z. Duan, J. Zhu and X. Lu, Discovery and Optimization of WDR5 Inhibitors *via* Cascade Deoxyribonucleic Acid-Encoded Library Selection Approach, *J. Med. Chem.*, 2024, **67**, 1079–1092.
69. G. Zhao, X. Fan, Y. Li, G. Zhang and Y. Li, Switchable DNA-Encoded Chemical Library: Interconversion between Double- and Single-Stranded DNA Formats, *ChemBioChem*, 2022, **23**, e202200025.
70. V. Blay, X. Li, J. Gerlach, F. Urbina and S. Ekins, Combining DELs and machine learning for toxicology prediction, *Drug Discovery Today*, 2022, **27**, 103351.
71. L. Prati, M. Bigatti, E. J. Donckele, D. Neri and F. Samain, On-DNA hit validation methodologies for ligands identified from DNA-encoded chemical libraries, *Biochem. Biophys. Res. Commun.*, 2020, **533**, 235–240.
72. J. Hall, T. L. Foley, Q. Chen, D. I. Israel, Y. Xu, K. F. Ford, P. Xie, J. Fan and J. Wan, A simple method for determining compound affinity and chemical yield from DNA-encoded library selections, *Biochem. Biophys. Res. Commun.*, 2020, **527**, 250–256.
73. J. C. Faver, K. Riehle, D. R. Lancia Jr., J. B. J. Milbank, C. S. Kollmann, N. Simmons, Z. Yu and M. M. Matzuk, Quantitative Comparison of Enrichment from DNA-Encoded Chemical Library Selections, *ACS Comb. Sci.*, 2019, **21**, 75–82.

CHAPTER 6

DNA-encoded Libraries with Covalent Warheads

SIXIU LIU,[a,b] RUI JIN,[a] ZHIQIANG DUAN[a] AND XIAOJIE LU*[a,b]

[a] State Key Laboratory of Drug Research, Shanghai Institute of Materia Medica, Chinese Academy of Sciences, Shanghai, 201203, China; [b] University of Chinese Academy of Sciences, No. 19A Yuquan Road, Beijing, 100049, China
*Email: xjlu@simm.ac.cn

6.1 Introduction

DNA-encoded library (DEL) technology provides an efficient hit identification platform for small-molecule drug discovery through affinity binding, and it has been broadly applied in academia and industry. DEL technology has been used in the field of compound screening to find reversible small molecules for 30 years and successfully discovered many small-molecule inhibitors. Several of these inhibitors have even entered the late-stage clinical studies.[1] For example in 2016,[2] GSK reported that they screened GSK's DNA-encoded small molecule libraries against RIP1 and identified a novel highly potent benzoxazepinone inhibitor series, which makes it an excellent starting point for further optimization into a RIP1 clinical candidate.

6.1.1 Covalent Inhibitors in Pharmacy

In recent years, an increasing number of covalent inhibitors, including FDA-approved ones such as Bruton's tyrosine kinase (BTK) inhibitor, ibrutinib, and epidermal growth factor receptor (EGFR) inhibitor, osimertinib, have seen significant sales. Ibrutinib had total sales of $8.43 billion in 2020, while

Drug Discovery Series No. 85
DNA-encoded Library Technology for Drug Discovery
Edited by Guansai Liu, Casey J. Krusemark and Jin Li

Published by the Royal Society of Chemistry, www.rsc.org

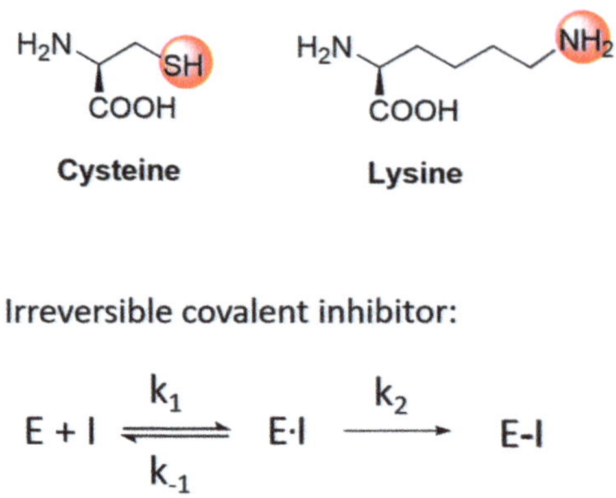

Figure 6.1 The interaction between a covalent inhibitor and common covalent-target amino acids.

osimertinib achieved sales of \$4.33 billion that same year.[3] Additionally, covalent modification enables effective inhibition targeting traditionally "untargetable" proteins. For instance, Amgen's Sotorasib, an inhibitor that covalently binds to KRAS G12C, has challenged the long-standing notion of KRAS protein being "undruggable".[4] The clinical success of these covalent inhibitors has garnered significant attention from scientists, propelling the development of small molecule covalent drugs. In comparison to non-covalent inhibitors, the interaction between a covalent inhibitor and its target protein involves reversible non-covalent interactions as well as covalent binding. Ideally, these interactions occur in two distinct steps (Figure 6.1). Usually, the first step is much faster than the second step. The reactivity of covalent inhibitors is time-dependent, as the accumulation of covalent binding over time leads to a higher ligand occupancy on the target protein. Based on the mode of action (MOA), covalent inhibitors can provide many pharmacological advantages, such as enhanced potency, selectivity, and prolonged duration of action.[5] Currently, clinically used covalent drugs primarily target critical cysteines,[1] with lysine,[6] serine,[7] and threonine[8] residues being targeted to a lesser extent.

6.1.2 Different Discovery Methods of Covalent Inhibitors

Historically, the discovery of most covalent inhibitors was accidental, and they were primarily selected from large-scale screening trials, which were not specifically designed for the covalent inhibitor discovery. Creating substrate mimics with electrophilic moieties based on catalytic mechanisms is the earliest rational design strategy,[9] to create so-called "suicide" inhibitors. Targeted covalent inhibitors, on the other hand, are not mechanism based and involve placement of a low-reactivity electrophile on a reversible small molecule inhibitor that creates a covalent linkage to a suitable protein nucleophile by induced proximity. Now, development methods of these covalent inhibitors are mainly divided into two major categories: ligand-first and electrophile-first approaches.[10] The ligand-first approach involves a structure-guided design by incorporating the "electrophilic functional group" into a known reversible ligand to enhance the inhibition of protein function. The electrophile-first approach is another increasingly used method for ligand discovery, in which covalent ligands are identified from the outset, rather than the combination of a covalent warhead with a known reversible ligand. All the above-mentioned discovery approaches of covalent inhibitors are summarized in Figure 6.2A–D.

Now, there are also other emerging technologies for covalent inhibitor discovery, such as intact protein MS,[11] covalent docking,[12] chemoproteomics-enabled discovery,[13] activity based protein profiling,[14] and covalent DNA encoded libraries.[15] A DEL with all compounds bearing an electrophilic warhead is defined as a covalent DEL, which is an electrophile-first approach (Figure 6.2). The covalent DNA-encoded library (covalent DEL, or CoDEL) provides an alternative method for screening small-molecule libraries of large sizes. CoDELs enable the screening of large covalent libraries through solid-phase protein immobilization, binder enrichment, DNA tag amplification, and DNA sequencing. Due to the irreversible interaction conferred by covalent bonds compared to non-covalent interactions, selection of covalent DEL molecules offers advantages over conventional affinity-based methods, allowing for higher signal-to-noise ratios in screening results through one single round of affinity selection. Compared with other methods, CoDEL enables the screening of massive libraries of small molecules, and it is time and cost efficient.

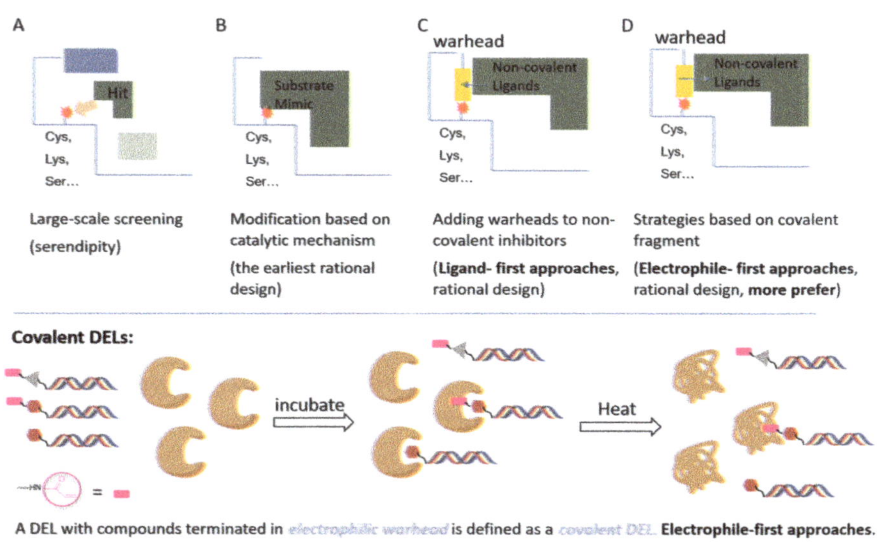

Figure 6.2 Different discovery approaches of covalent inhibitors.

6.2 Hits Selected from CoDELs

Now, CoDELs have been screened against a wide range of protein targets, including BTK,[16,17] JAK1,[17] Pin1,[17] and 3CLpro of SARS-CoV-2,[18] resulting in the identification of unique covalent ligands. In 2017, the David Liu group[19] reported the discovery of a covalent inhibitor against MAP2K6 kinase through the combination of a DNA-encoded library and the Interaction Determination Using Unpurified Proteins (IDUP) method. In the same year, Neri's group[20] reported the discovery of specific covalent small-molecule inhibitors for JNK1 by generating two self-assembling libraries consisting of 265 and 559 members. As early as 2015, X-Chem presented its covalent DEL design and the screening results against the BTK target at the 1st Boston Symposium of Encoded Library Platform. Furthermore, in 2021, they reported[16] the discovery of a covalent inhibitor for the BTK kinase and the corresponding screening method. By optimizing the screening conditions, they discovered several new series of covalent irreversible inhibitors for BTK that interact with cysteine. In 2022, WuXi AppTec[18] used conventional acylation reactions to construct a covalent DNA-encoded peptide library with an acrylamide warhead. Through covalent screening, they obtained novel lead compounds from two series. Subsequent structure optimization led to the identification of the most active compound, which demonstrated inhibition activity against the 3CL protein with an approximate 2 μM IC_{50}. Additionally, our group[17] designed several DNA-encoded libraries of triazine-based scaffolds. After validating the covalent selection method using spebrutinib as a positive on-DNA molecule, we successfully conducted covalent screening against the targets BTK, JAK3, and Pin1, resulting in the discovery of several novel covalent inhibitors. The representative covalent inhibitors selected from CoDELs are listed in Table 6.1.

Currently, most covalent small molecules discovered through covalent CoDELs are designed to target proteins by specifically reacting with cysteine residues that are conserved in their structure. Cysteine residues are often preferred in covalent drug design due to their strong nucleophilic nature, which allows them to form stable covalent bonds with warhead groups in these molecules. By exploiting this reactivity, CoDELs can selectively bind to these cysteine sites, thereby enhancing the specificity and efficacy of the small molecule in targeting proteins associated with disease processes. Although cysteine has the most profound nucleophilic character, its labeling potential for clinically important protein targets is limited due to its relatively low abundance (2.3%) in the human proteome.[21] Additionally, many relevant drug targets lack reactive and accessible cysteine residues at tractable binding sites. The presence of disulfide bridges among cysteines hinders their covalent modification. Because of these challenges, an alternative covalent conjugation strategy becomes essential when cysteine or other reactive residues are not available in the protein of interest. Therefore, the synthesis of CoDELs that can target multiple amino acid residues is a promising strategy for extending the utility of covalent inhibitors and an active area of research.

Table 6.1 Covalent hits selected from CoDELs

Target	DEL hit molecule	Properties
MAP2K6[19]	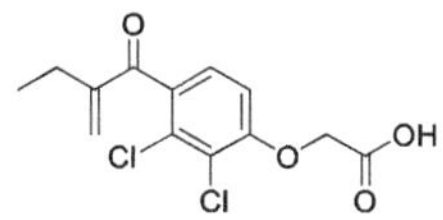	MW: 303.14 $IC_{50} = 4.5$ μM
JNK1[20]	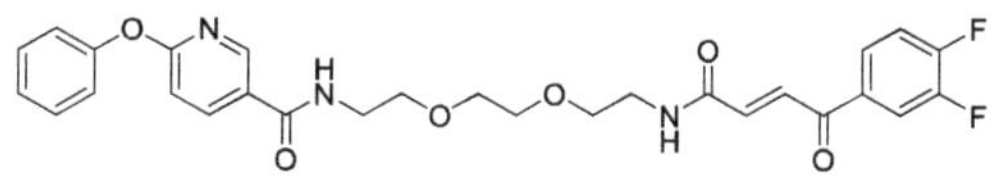	MW: 539.54
BTK[16]	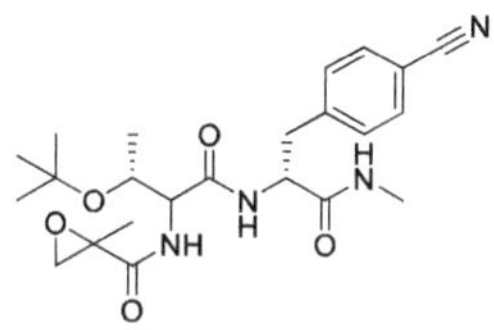	MW: 444.53 $IC_{50} = 58$ nM
SARS Cov-2,3 CLpro[18]	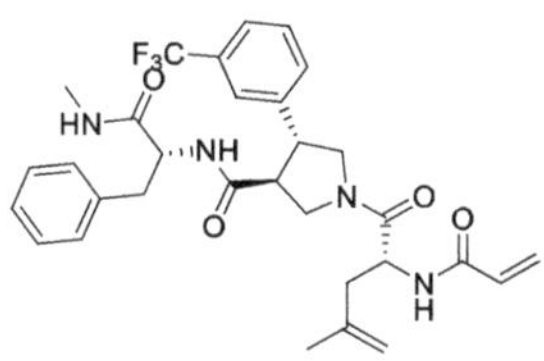	MW: 584.64 $IC_{50} = 2.0$ μM
JAK3[17]	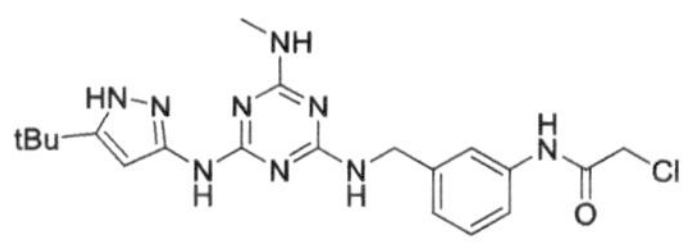	MW: 584.64 $IC_{50} = 6.16$ nM
Pin1[17]	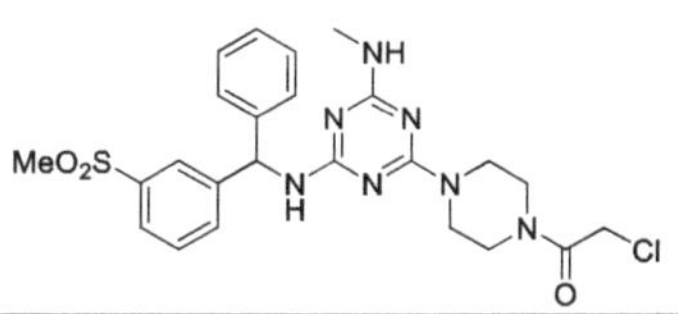	MW: 530.04 $IC_{50} = 1.69$ μM

Recently, our group, in collaboration with Professor Zhou and Tan,[22] reported the discovery of covalent ligands that target tyrosine residues from sulfonyl fluoride-containing CoDELs. In this project, researchers established a high-throughput screening platform that integrated activity-based protein profiling (ABPP) using sulfonyl fluoride-based covalent probes with the sulfonyl fluoride CoDELs, which provides a generalizable workflow for discovering tyrosine-targeting covalent ligands (Figure 6.3). We chose PGAM1, GSTP1 and DPP3 as targets with tyrosine reactive hotspots for sulfonyl fluoride probes as demonstrated by ABPP results. Through screening a DEL pool containing 67 million compounds, we identified novel covalent ligands that selectively target tyrosine residues in PGAM1, GSTP1, and DPP3, respectively. The covalent inhibitors of these three targets have not been thoroughly investigated. Among them, DPP3 was the least studied whose

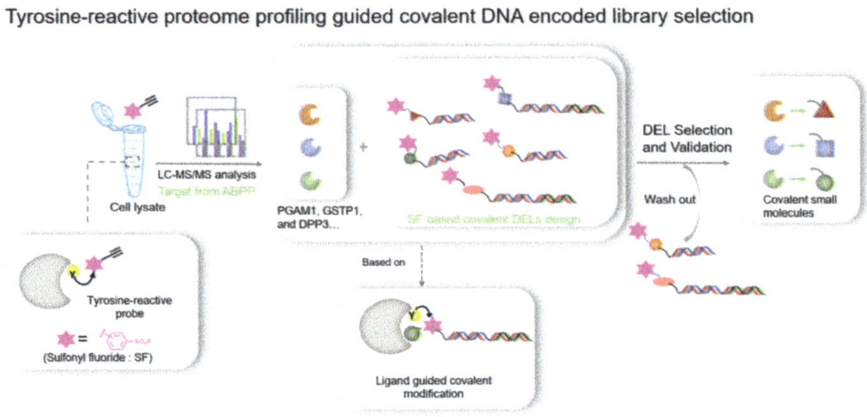

Figure 6.3 ABPP-guided discovery of tyrosine-targeting covalent inhibitors from DELs.

covalent inhibitors had not been previously reported. Development of DPP3 inhibitors itself is challenging due to the extended substrate binding pocket. This demonstration highlights the potential of sulfonyl fluoride as a promising warhead for tyrosine-targeting covalent ligands and emphasizes the importance of combining ABPP and CoDELs for the rapid discovery of covalent inhibitors. While this method successfully led to the discovery of covalent ligands, it still requires further extension and development. One key issue is the limited number of core scaffolds and the relatively small chemical space in CoDELs, which may constrain the selectivity of hits generated by this method.

6.3 Warheads and On-DNA Chemistry in CoDELs

6.3.1 Reported Warheads Targeting Cysteine Residues and On-DNA Chemistry

Based on the earlier description, it is currently observed that covalent inhibitors identified through the DEL technique primarily target cysteine residues. Cysteines are relatively rare in the human proteome, but they have been found to react with a variety of warheads, particularly Michael acceptors.[23,24] Extensive research has been conducted to target cysteine residues, leading to the exploration of a wide variety of covalent warheads in the study of small-molecule covalent inhibitors. The majority of reported covalent warheads in DELs are α,β-unsaturated ketone Michael acceptors, such as acrylamide derivatives. Figure 6.4 presents a summary of the covalent warheads and commercially available building blocks that are used in the reported DELs in the current literature. The covalent group of acrylamide forms irreversible conjugates with the thiol group of cysteine through Michael addition, which is an effective method to

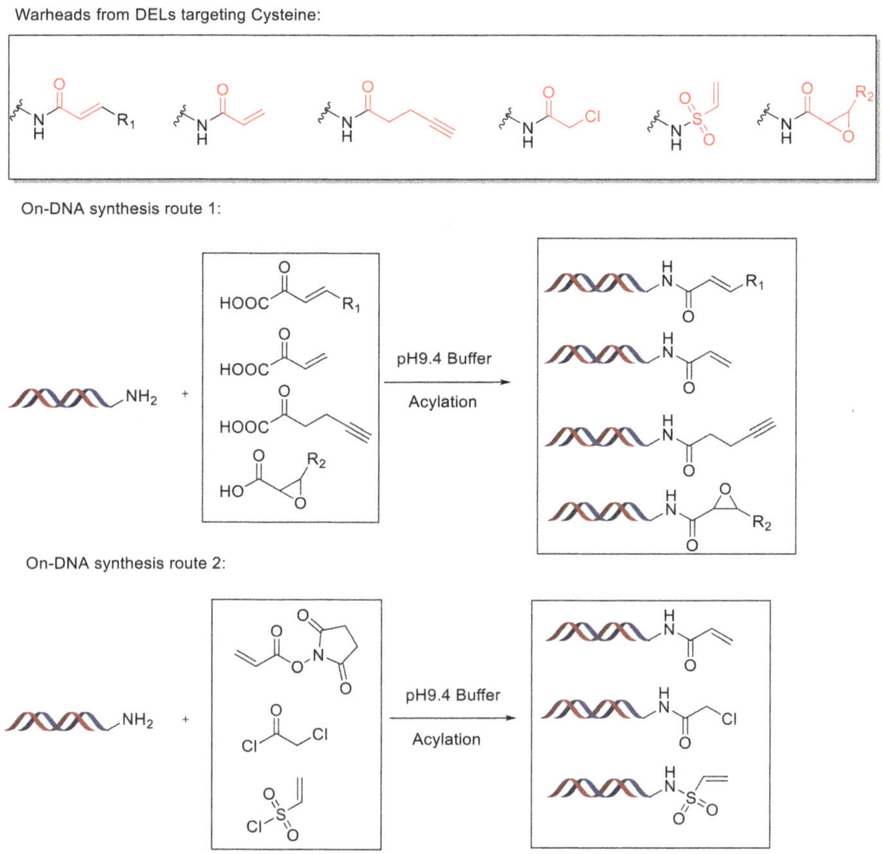

Figure 6.4 Warheads and on-DNA chemistry in CoDELs.

enhance the potency and selectivity of irreversible inhibitors.[25,26] In addition to acrylamide, other alkenes and alkynes can also undergo Michael addition with thiols. α-Halo acetamides are another well-defined class of cysteine-directed electrophiles that react with thiols *via* an S_N2 mechanism.[27]

Before constructing a covalent DEL, the stability of the covalent warheads on the DNA needs to be tested. This includes evaluating warhead tolerance to chemical reaction conditions, ligation conditions, and selection conditions. Generally, the chemical reaction used to connect the warhead to the DEL involves the formation of an amide bond. Additionally, nucleophilic substitution reactions utilizing sulfonyl chloride reagents to form sulfonamide are also employed (Figure 6.4). After conjugating the warheads, the on-DNA stability can be analyzed by molecular weight (MW) changes indicated by LC–MS. If the potential reaction with DNA cannot be reflected on the MW changes, PCR or Sanger sequencing can be employed for further validation.

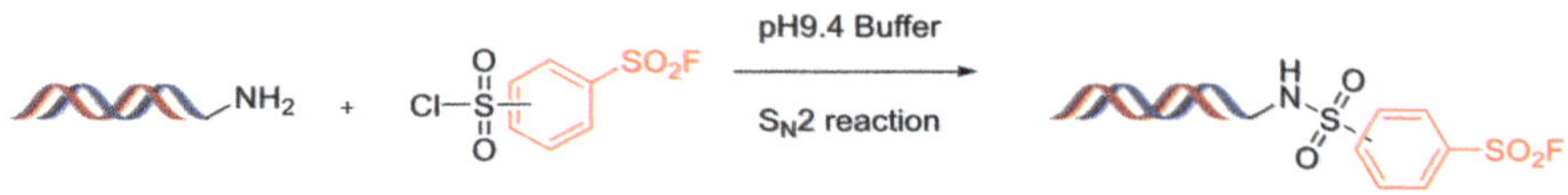

Scheme 6.1 On-DNA sulfonamide formation to install a reactive sulfonyl fluoride warhead.

6.3.2 Reported Warheads Targeting Non-cysteine Residues and On-DNA Chemistry

As discussed before, the majority of covalent inhibitors discovered using covalent DEL technology react with conserved cysteine residues within or adjacent to the ligand binding site. This limits its further application since many target proteins lack suitable cysteine residues. Therefore, research on covalent DELs targeting non-cysteine residues is an essential direction to pursue. Recently, Professor Yizhou Li[28] reported a covalent selection method *via* photo-activated lysine-selective crosslinkers. Above, there is a mention of our collaboration work on tyrosine targeting, in which CoDEL was combined with the sulfonyl fluoride-based ABPP technique.[22] Compared to traditional methods of screening covalent inhibitors, our approach offers higher efficiency, better stability, and improved specificity. Traditional screening of covalent inhibitors often involves arbitrary selection of targets or covalent warhead types. In contrast, our method allows for the rational design and selection of covalent warheads and target proteins, thereby enabling the discovery of specific covalent inhibitors. This demonstration emphasizes the importance of combining ABPP and CoDEL for the rapid discovery of covalent inhibitors and highlights the potential of sulfonyl fluoride as a promising warhead for tyrosine-targeting covalent ligands. The chemical reaction used to attach the sulfonyl fluoride warhead to the DEL involves nucleophilic substitution reactions as shown in Scheme 6.1.

6.3.3 Potential Warheads and On-DNA Chemistry in CoDELs

Distinct from the requirements for covalent warheads in small molecule inhibitors, all covalent warheads used in DELs need to satisfy certain conditions, including stability and reactivity in aqueous solvents, compatibility with enzyme-ligation reactions and affinity-based selection. Currently, structurally simple covalent warheads are used in DELs, limiting their applicability for target screening. Therefore, more efforts should be devoted to exploring novel covalent warheads for DELs. In fact, many covalent warheads have been reported for small-molecule inhibitors. Some of them may be potentially useful for constructing CoDELs. Here, we introduce covalent warheads that target cysteine residues and non-cysteine residues. Covalent warheads targeting cysteine residues include α,β-unsaturated ketones as discussed previously, as well as other chemical fragments such as heterocyclic aromatics,[29] sulfonyl pyridines,[30] cyanides,[31] bridged rings,[32]

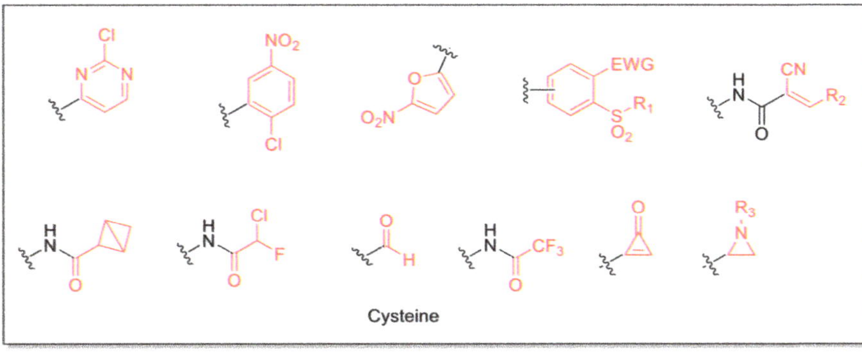

Figure 6.5 Potential warheads targeting cysteine for CoDELs.

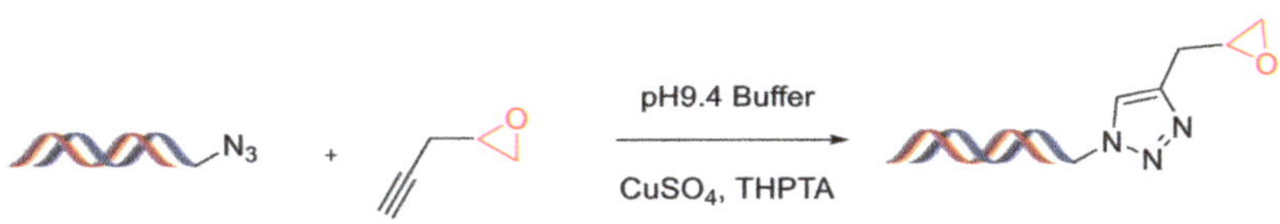

Scheme 6.2 On-DNA click reaction to install a reactive epoxide warhead.

α-chlorofluoroacetamides,[33,34] aldehydes,[35–37] trifluoromethyl ketones,[38] and aziridines[39] (Figure 6.5).

Some of these covalent warheads have been thoroughly studied, validated, and even applied in the clinical therapy. We also need to conduct stability tests to determine how these warheads can be incorporated in DELs. Heterocyclic aromatics as S_NAr warheads through a stepwise addition–elimination mechanism have been used in covalent inhibitors.[29] These include 2-sulfonylpyridines as tuneable S_NAr-based reactive groups selectively react with thiols *via* nucleophilic aromatic substitution. Covalent warheads in DELs targeting cysteine residues can be classified into two major categories based on their reaction types: nucleophilic substitutions (S_N) and the Michael-type additions (Ad_{NM}). The S_N type includes halogenated alkyls, epoxides, and other three-membered rings. The Ad_{NM} type includes acrylamide and alkynyl groups as mentioned earlier.[40]

In DEL construction, building blocks (BBs) bearing these covalent warheads commonly undergo amide bond formation and nucleophilic substitution reactions for DNA conjugation. However, for reagents that are difficult to purchase or synthesize, one can consider using covalent BBs with synthetic alkynyl groups and then employ click chemistry for the covalent linkage between the reactive group moiety and the DNA tag. This approach benefits from the strong functional group compatibility and widespread application of click reactions in chemical biology. The standard on-DNA procedure for click reactions involves the utilization of copper-catalyzed azide–alkyne cycloaddition (CuAAC) as shown in Scheme 6.2.[41]

As mentioned before, research on covalent libraries is still in its early stages, and there has not been an extensive investigation into CoDELs targeting non-cysteine residues. However, significant progress has been made in the development of small molecule covalent inhibitors that target non-cysteine residues. We can refer to the reported small molecule covalent warheads for DEL conjugation and conduct further research on CoDELs based on these findings.

In recent years, lysine residues have gained attention for their nucleophilic role in covalent labeling. Lysine is one of the main targets for covalent warheads due to its ubiquitous presence in human proteins.[42] Through the investigation of small-molecule covalent binders on different proteins, we have identified several covalent warheads that are likely to find applications in CoDELs. While numerous lysine-targeting warheads have been reported in the literature,[43] further validation is required to ensure their stability. We have only listed the covalent warheads that are deemed feasible for synthesis in DELs, along with proposing the conjugation method to their corresponding linker DNA (Figure 6.6). Like BBs containing covalent warheads targeting cysteines, the chemistry involved in conjugating lysine-targeting BBs to DNA typically includes amide bond formation and nucleophilic substitution reactions. Lysine-targeting warheads generally adopt similar mechanisms of covalent interaction with lysine to that with cysteine. However, among these mechanisms, the covalent interaction of the *o*-nitrobenzyl

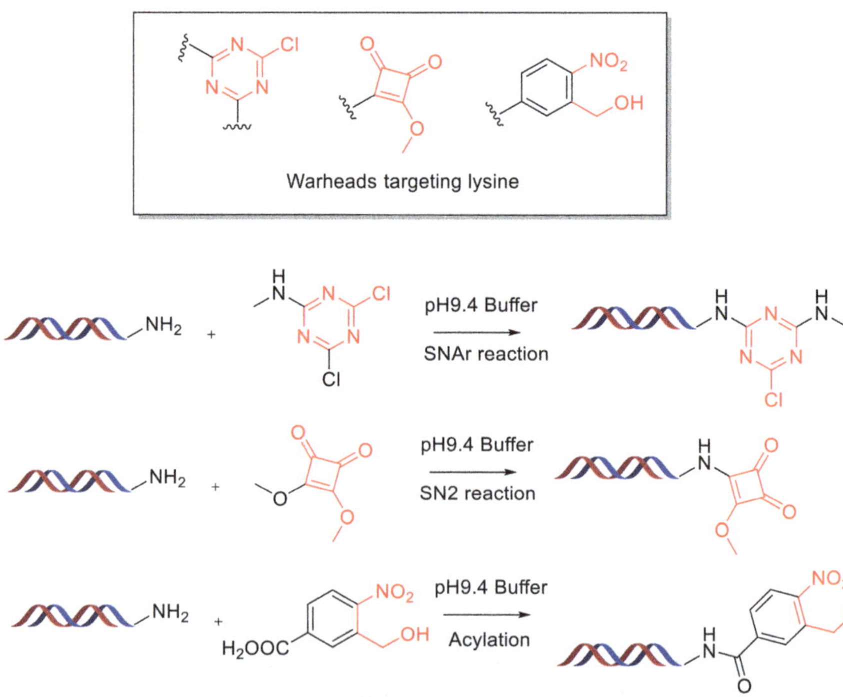

Figure 6.6 Potential warheads targeting lysine for CoDELs and suitable conjugation chemistries.

alcohol (*o*-NBA) warhead with lysine differs from other warheads as it involves photo-promoted selective covalent interaction with lysine.[44] Therefore, DELs utilizing this specific covalent warhead differ from other covalent libraries in terms of the covalent selection method. The photocatalytic selection method should be applied, which was recently reported by Professor Yizhou Li.[28]

When it comes to covalent warheads targeting other amino acid residues, there are relatively fewer reports available. However, covalent warheads such as sulfonyl fluoride and Mannich electrophiles,[45] which target tyrosine, remain to be potential research subjects. While we have reported on the application of aromatic sulfonyl fluoride in CoDELs, aliphatic sulfonyl fluoride covalent warheads still present opportunities for further investigation.[43] Recently, the Krusemark group[45] reported on novel cyclic imine warheads and demonstrated the first application of ligand-directed Mannich chemistry for tyrosine conjugation. These findings also suggest that these warheads have the potential to be used in CoDELs.

6.4 Covalent Selection Methods

Due to the increased stability of covalent interactions compared to non-covalent interactions, typically only one round of selection is needed to achieve a favorable signal-to-noise ratio in the conventional covalent selection method. The selection method for covalent DELs typically involves a four-step process. First, the library is incubated with the target of interest in solution, allowing compounds to bind covalently to their targets. The second step is to immobilize this mixture with beads to capture the target protein and the associated DEL molecules. Then, the unbound compounds are washed away, followed by heat treatment, in which the protein is denatured allowing removal of any reversible binders. Next, the DNA tag of irreversible binders remains immobilized on the affinity matrix, which are directly amplified by PCR for sequencing.[44] Covalent protocols published by Wuxi AppTec[18] and X-Chem[16] apply the same principles in their methods. However, there are still some differences in specific details among them. Wuxi AppTec states that DTT and TCEP are typically excluded from the incubation buffer to avoid potential reactions with electrophiles in covalent libraries. In X-Chem's protocol, they perform heat denaturation first followed by the immobilization of DEL molecules and protein on beads. Both the Wuxi and X-Chem protocols involve eluting the complexes of DEL molecules with target proteins from the matrix using a high concentration of imidazole when his-tagged proteins are used for immobilization. This indicates that they employ an elution sample without beads for qPCR and PCR purposes.

6.5 CoDEL Synthesis

In the field of CoDEL synthesis, there is currently limited availability of intermediate libraries that can be connected to covalent warheads. It is important to design and synthesize a diverse set of intermediate libraries with chemically spacious heterocyclic scaffolds, which can be connected to covalent warheads through DNA-compatible chemical reactions. The typical design of our intermediate library includes exposed amine groups that are capable of undergoing reactions such as amide bond condensation and nucleophilic substitution with BBs bearing covalent warheads. Additionally, primary amines can be converted to azides through functional group modifications, enabling click reactions with alkyne group BBs. Furthermore, it is possible to synthesize libraries of intermediates with other exposed groups, such as the aldehyde group, which serves as a reversible covalent connector targeting cysteine. Dr Yunjin Hu recently published[46] research on the Morita–Baylis–Hillman (MBH) reaction on DNA, demonstrating its ability to convert the aldehyde group into an α,β-Michael acceptor. This outcome presents its potential as a covalent warhead in the development of covalent inhibitors.

After identifying promising lead compounds for certain target of interest (which can be reported compounds), a focused covalent DEL can be designed based on the structure types of these compounds. This approach leverages the advantages of DNA-encoded libraries to expand chemical diversity while maintaining reasonable compound activity. It enables rapid exploration of structure–activity relationships and facilitates further optimization of compound structures.

6.6 Conclusions and Future Perspectives

Covalent DNA-encoded libraries have gained significant attention for their use in drug discovery and chemical biology research. In these libraries, covalent warheads play a crucial role. In this context, we provide an overview of the latest progress in covalent warhead development for DELs. The current research efforts are primarily directed towards synthesizing diverse DNA-encoded libraries and creating innovative and efficient on-DNA chemistry methods to enhance the stability and reactivity of the conjugated covalent warheads. Various strategies have been conducted to introduce covalent warheads, such as amide-bond formation and nucleophilic chemical reactions. Additionally, advances in synthetic chemistry have facilitated the discovery of novel on-DNA chemistries and more efficient coupling methods. However, despite these notable improvements, several technical challenges and limitations remain. One significant limitation is the lack of stable covalent warheads, as many of the current warheads suffer from low conversion rates and unstable reaction intermediates. Moreover, designing and synthesizing covalent warheads continue to pose significant challenges due to the intricate nature of DNA tags and the requirement for compatibility with diverse chemical functional groups. To overcome these challenges, future directions in covalent warhead design may involve the advancement of new on-DNA chemistries and the implementation of more efficient coupling methods. Additionally, there is a necessity for an enhanced understanding of the factors influencing covalent bond formation between the compound and the DNA tag. Overall, covalent warhead design is an active and rapidly evolving area of research in DELs. Progress in this area will facilitate the creation of more stable and effective covalent warheads, thereby enhancing hit identification accuracy and the potential for drug candidate discovery.

In addition to constructing diverse DELs, the development of novel covalent warheads is also crucial for the study of covalent inhibitors. In addition to targeting cysteine and tyrosine residues, the development of covalent inhibitors can be extended to other amino acid residues. The construction of CoDELs for covalent screening involves the use of covalent warheads capable of connecting with DNA-encoded libraries. It is crucial to consider factors such as the stability of the covalent warhead, compatibility with DNA chemical reactions, and stability under screening conditions when extending the use of other covalent warheads in conjunction with DNA-encoded libraries. This is to ensure seamless progress of the screening process and to minimize the occurrence of false positive signals. Currently, there are 9 amino acids (His, Arg, Lys, Met, Cys, Asp, Glu, Tyr, and Asn) reported to have the capability of engaging in covalent interactions.[18] By combining these reported covalent warheads with newly developed novel ones within DNA-encoded libraries, it is expected that this approach will enable the discovery of covalent inhibitors that specifically target proteins containing various amino acid residues.

The covalent warhead in DELs offers significant advantages and promising applications in drug discovery and biological research. Its importance lies in the enhanced stability and specificity it provides, leading to more accurate hit identification during screening processes. This contributes to the development of potential drug candidates and the identification of compounds that specifically target proteins with reactive residues.

The prospects for covalent warheads in DELs are exceedingly encouraging. The role of CoDELs in expediting the discovery of novel drug candidates and promoting chemical biology research is paramount. Further research and exploration of the potential of covalent linkers in DELs is strongly encouraged, as their continued progress holds great promise for the field of drug discovery and biological studies. Valuable research directions for CoDEL development can focus on two areas: firstly, enriching the diversity of DELs to expand chemical diversity, and secondly, increasing the diversity of covalent warheads (which can involve using well-established stable covalent warheads as well as developing novel covalent warheads). Moreover, by conjugating other biochemical methods like ABPP, we are able to apply CoDELs to protein targets that have a higher potential to interact with certain covalent warheads. We are confident that recent advancements and ongoing research into CoDELs will create additional opportunities for discovering covalent inhibitors that target both cysteine and non-cysteine residues.

References

1. P. R. Fitzgerald and B. M. Paegel, *Chem. Rev.*, 2021, **121**, 7155.

2. P. A. Harris, B. W. King and D. Bandyopadhyay, *et al.*, *J. Med. Chem.*, 2016, **59**, 2163.
3. E. Sagonowsky, The Top 20 Drugs by Worldwide Sales in 2020, Fierce Pharma, 2021, https://www.fiercepharma.com/special-report/top-20-drugs-by-2020-sales.
4. A. R. Moore, S. C. Rosenberg, F. McCormick and S. Malek, *Nat. Rev. Drug Discovery*, 2020, **19**, 533.
5. T. A. Baillie, *Angew. Chem., Int. Ed.*, 2016, **55**, 13408.
6. D. Quach, G. Tang, J. Anantharajan and N. Baburajendran, *et al.*, *Angew. Chem., Int. Ed.*, 2021, **60**, 17131.
7. O. O. Fadeyi, L. R. Hoth, C. Choi, X. Feng, A. Gopalsamy, E. C. Hett, R. E. Kyne, R. P. Robinson and L. H. Jones, *ACS Chem. Biol.*, 2017, **12**, 2015.
8. A. T. Bender, A. Gardberg, A. Pereira, T. Johnson, Y. Wu, R. Grenningloh, J. Head, F. Morandi, P. Haselmayer and L. Liu-Bujalski, *Mol. Pharmacol.*, 2017, **91**, 208.
9. S. De Cesco, J. Kurian, C. Dufresne, A. K. Mittermaier and N. Moitessier, *Eur. J. Med. Chem.*, 2017, **138**, 96.
10. L. Boike, N. J. Henning and D. K. Nomura, *Nat. Rev. Drug Discovery*, 2022, **21**, 881.
11. E. Resnick, A. Bradley, J. Gan, A. Douangamath, T. Krojer and R. Sethi, *et al.*, *J. Am. Chem. Soc.*, 2019, **141**, 8951.
12. A. Scarpino, G. G. Ferenczy and G. M. Keserű, *J. Chem. Inf. Model.*, 2018, **58**, 1441.
13. B. F. Cravatt, A. T. Wright and J. W. Kozarich, *Annu. Rev. Biochem.*, 2008, **77**, 383.
14. K. M. Backus, B. E. Correia and K. M. Lum, *et al.*, *Nature*, 2016, **534**, 570.
15. S. Brenner and R. A. Lerner, *Proc. Natl. Acad. Sci. U. S. A.*, 1992, **89**, 5381.
16. J. P. Guilinger, A. Archna, M. Augustin, A. Bergmann, P. A. Centrella and M. A. Clark, *et al.*, *Bioorg. Med. Chem.*, 2021, **42**, 116223.
17. L. Li, M. Su, W. Lu, H. Song, J. Liu, X. Wen, Y. Suo, J. Qi, X. Luo, Y. Zhou, X. H. Liao, J. Li and X. Lu, *ACS Med. Chem. Lett.*, 2022, **13**, 1574.
18. R. Ge, Z. Shen, J. Yin, W. Chen, Q. Zhang, Y. An, D. Tang, A. L. Satz, W. Su and L. Kuai, *SLAS Discovery*, 2022, **27**, 79.
19. A. I. Chan, L. M. McGregor, T. Jain and D. R. Liu, *J. Am. Chem. Soc.*, 2017, **139**, 10192.
20. G. Zimmermann, U. Rieder, D. Bajic, S. Vanetti, A. Chaikuad, S. Knapp, J. Scheuermann, M. Mattarella and D. Neri, *Eur. J. Chem.*, 2017, **23**, 8152.
21. N. Péczka, *et al.*, *Expert Opin. Drug Discovery*, 2022, **17**, 413.
22. L. Jiang, S. Liu, X. Jia, Q. Gong, X. Wen, W. Lu, J. Yang, X. Wu, X. Wang, Y. Suo, Y. Li, M. Uesugi, Z. Qu, M. Tan, X. Lu and L. Zhou, *J. Am. Chem. Soc.*, 2023, **145**, 25283.
23. S. De Cesco, J. Kurian, C. Dufresne, A. K. Mittermaier and N. Moitessier, *Eur. J. Med. Chem.*, 2017, **138**, 96.
24. P. A. Jackson, J. C. Widen, D. A. Harki and K. M. Brummond, *J. Med. Chem.*, 2017, **60**, 839.

25. I. M. Serafimova, M. A. Pufall, S. Krishnan, K. Duda, M. S. Cohen, R. L. Maglathlin, J. M. McFarland, R. M. Miller, M. Frödin and J. Taunton, *Nat. Chem. Biol.*, 2012, **8**, 471.
26. H. Kim, Y. S. Hwang, M. Kim and S. B. Park, *RSC Med. Chem.*, 2021, **12**, 1037.
27. R. Ahmadi and S. Emami, *Eur. J. Med. Chem.*, 2022, **234**, 114255.
28. H. Wei, T. Zhang, Y. Li, G. Zhang and Y. Li, *Chem. – Asian. J.*, 2023, **18**, e202300652.
29. M. Gehringer and S. A. Laufer, *J. Med. Chem.*, 2019, **62**, 5673.
30. C. Zambaldo, E. V. Vinogradova, X. Qi, J. Iaconelli, R. M. Suciu, M. Koh, K. Senkane, S. R. Chadwick, B. B. Sanchez and J. S. Chen, *et al.*, *J. Am. Chem. Soc.*, 2020, **142**, 8972.
31. R. M. Miller, V. O. Paavilainen, S. Krishnan, I. M. Serafimova and J. Taunton, *J. Am. Chem. Soc.*, 2013, **135**, 5298.
32. K. Tokunaga, M. Sato, K. Kuwata, C. Miura, H. Fuchida, N. Matsunaga, S. Koyanagi, S. Ohdo, N. Shindo and A. Ojida, *J. Am. Chem. Soc.*, 2020, **142**, 18522.
33. N. Shindo, H. Fuchida, M. Sato, K. Watari, T. Shibata, K. Kuwata, C. Miura, K. Okamoto, Y. Hatsuyama and K. Tokunaga, *et al.*, *Nat. Chem. Biol.*, 2019, **15**, 250.
34. M. Sato, H. Fuchida, N. Shindo, K. Kuwata, K. Tokunaga, G. Xiao-Lin, R. Inamori, K. Hosokawa, K. Watari and T. Shibata, *et al.*, *ACS Med. Chem. Lett.*, 2020, **11**, 1137.
35. J. C. Powers, J. L. Asgian, Ö. D. Ekici and K. E. James, *Chem. Rev.*, 2002, **102**, 4639.
36. M. C. Noe and A. M. Gilbert, in *Annual Reports in Medicinal Chemistry*, Academic Press, Cambridge, 2012.
37. M. Siklos, M. BenAissa and G. R. J. Thatcher, *Acta Pharm. Sin. B*, 2015, **5**, 506.
38. Z. Zhang, Y. Wang, X. Chen, X. Song, Z. Tu, Y. Chen, Z. Zhang and K. Ding, *Bioorg. Med. Chem.*, 2021, **50**, 116457.
39. G. S. Singh, *Mini-Rev. Med. Chem.*, 2016, **16**, 892.
40. F. Huang, X. Han, X. Xiao and J. Zhou, *Molecules*, 2022, **27**, 7728.
41. X. Ling, S. Liu, Y. Yang, Q. Dong, L. A. Marcaurelle, W. Huang, Y. Ding, X. Wang and X. Lu, *Bioconjugate Chem.*, 2023, **34**, 616.
42. L. H. Jones, *Annu. Rep. Med. Chem.*, 2021, **56**, 95.
43. M. E. Abbasov, M. E. Kavanag, T. Ichu, M. R. Lazear, Y. Tao and V. M. Crowley, *et al.*, *Nat. Chem.*, 2021, **13**, 1081.
44. A. D. Guo, *et al.*, *Nat. Commun.*, 2020, **11**, 5472.
45. S. Wang, M. Hadisurya, W. A. Tao, E. Dykhuizen and C. Krusemark, *ChemRxiv*, DOI: 10.26434/chemrxiv-2022-tvgn1.
46. L. Xue, W. Liu and S. Li, *et al.*, *Bioconjugate Chem.*, 2023, **34**, 1366.

CHAPTER 7

Application of DELs for E3 Ligase Ligand Discovery and Targeted Protein Degradation

MARIE L. MALONE, NICHOLAS A. SANCHEZ, STACY L. HU AND CHRISTOPHER B. PHELPS*

Nurix Therapeutics, 1700 Owens STE 205, San Francisco, 94158, USA
*Email: cphelps@nurixtx.com

7.1 Overview of Targeted Protein Degradation

First conceptualized by Crews and Deshaies in the early 2000s, targeted protein degradation (TPD) represents a groundbreaking therapeutic strategy with the potential to target virtually any protein for therapeutic effect.[1] Small molecules employed in TPD strategies induce proximity between a protein of interest (POI) and the cell's native protein degradation machinery creating neo-interactions that lead to routing of the POI to the proteasome for elimination from the cell. The majority of therapeutics that have advanced into the clinic that employ this strategy function through small molecule recruitment of a POI to a non-native E3 ubiquitin ligase, resulting in target ubiquitylation and subsequent degradation by the ubiquitin proteosome system (UPS).[2] These are broadly categorized into two main types: bifunctional degraders (coined PROteolysis TArgeting Chimeras (PROTACs) by Crews and Deshaies), which utilize chemically linked binders specific for the target protein and an E3 ligase, and molecular glue degraders (MGDs) – monovalent compounds that induce cooperative binding between a target and an E3 ligase. Targeted protein degradation strategies may provide a

Drug Discovery Series No. 85
DNA-encoded Library Technology for Drug Discovery
Edited by Guansai Liu, Casey J. Krusemark and Jin Li

Published by the Royal Society of Chemistry, www.rsc.org

dramatic therapeutic advantage over direct enzymatic inhibition strategies for key scaffolding targets or instances where inhibiting enzymatic activity is insufficient for full pathway modulation.[3–8] In this chapter, we will highlight applications and advancements in DEL discovery for the development of bifunctional and molecular glue degraders.

Bifunctional degraders employ a design where binders for both the E3 ligase and the target are connected by a central linker. This design facilitates the formation of a ternary complex between the ligase, target, and ligand, leading to the target's ubiquitylation and subsequent degradation.[1] In the first published example, PROTAC-1 recruits the βTrCP/SKP1 E3 ligase complex to neo-substrate MetAP-2 for ubiquitylation and degradation in *Xenopus* oocyte extracts.[1] While bifunctional degraders generally have larger molecular weights (>700 Da) and physiochemical properties beyond the rule-of-five space, reports of oral bioavailability and central nervous system (CNS) penetration are becoming commonplace.[9–13] Numerous bifunctional degrader therapeutics have entered clinical trials in the areas of oncology and inflammation, examples of which include CRBN-based androgen receptor (AR) degraders for the treatment of prostate cancer (ARV-766), a CRBN-based estrogen receptor α degrader currently for the treatment of ER+/HER2− breast cancer (ARV-471), the CRBN-based IL-1 receptor-associated kinase 4 (IRAK4) degrader KT-474 for the treatment of hidradenitis suppurativa and atopic dermatitis, CRBN-based Bruton's tyrosine kinase (BTK) degraders for the treatment of B cell malignancies (NX-5948 and NX-2127), the von Hippel–Lindau (VHL)-based STAT3 degrader for treatment of hematological malignancies (KT-333), and the CRBN-based BRAF V600E degrader CFT1946 for the treatment of BRAF V600E mutant colorectal and non-small cell lung cancers.[2,14–23]

Molecular glue degraders (MGDs) are monovalent compounds that bind cooperatively to induce a PPI interface between a POI and E3 ubiquitin ligase, either by stabilizing a native low affinity interaction or by inducing a neomorphic interaction.[24,25] Binding at the interface of an E3 ligase–neo-substrate complex, MGDs can be classified by an increase in the affinity to the complex relative to its individual components.[26] In nature, MGDs can be found in plant hormone signaling such as the auxin or jasmonate pathways.[27–29] Elucidating the molecular mechanisms of these naturally occurring and serendipitously discovered MGD pathways has inspired many TPD development and therapeutic strategies in the last two decades. Well-defined examples of MGD drugs are the immunomodulatory drugs (IMiDs) derived from thalidomide, which are used in the treatment of hematologic malignancies.[30] IMiDs bind CRBN and modify its protein surface, enabling engagement of non-native substrates such as IKZF1/3 transcription factors, leading to their subsequent proteolytic degradation.[31–33] Interestingly, the IMiD pharmacophore resembles CRBN's physiological substrate, recently described to be a post-translational protein modification that arises from intramolecular cyclization of glutamine or asparagine residues.[34]

Targeted protein degradation strategies provide a means to broaden the reach of traditional drug discovery, expanding the breadth of druggable targets beyond enzyme-focused drug discovery into target classes with non-enzymatic functions, key scaffolding roles, or lesser defined interaction motifs.[3–8] Recent advances have also demonstrated that bifunctional degraders can show resilience to some forms of drug resistance, specifically amino acid point mutations that often drive resistance to enzymatic inhibitors used in oncology, demonstrating one of the potential advantages of degrader-based therapeutics.[3] The catalytic nature of targeted protein degraders, where each molecule can engage a ligase resulting in ubiquitylation and degradation of many copies of a single protein target, is another potential advantage and may provide opportunities for reducing drug exposure in patients.[2] The current repertoire of E3 ligases routinely coopted for TPD (CRBN, VHL, MDM2, DDB1, DCAF15, and SCF βTrCP) is but a small fraction of the more than 600 E3 ligases encoded by the human genome, underscoring the untapped therapeutic potential for targeting new UPS degradation pathways with novel ligase binders.

7.2 DELs for the Discovery of TPD Compounds

DNA-encoded library (DEL) technology has become a significant technique in the development of TPD therapeutics. The design of TPD compounds fundamentally does not depend on functionally relevant target protein binding pockets, but rather requires only an interaction between a POI and a ligase to achieve effective ubiquitylation.[35] Complimenting this approach, DEL screening methodology does not rely on the knowledge of or assays for a specific protein binding pocket, and thus it has the potential to identify binders to multiple pockets in a single screening effort (Figure 7.1A).[36] Depending on the therapeutic goals, the streamlined workflow of affinity-based DEL screening can be utilized to identify selectively targeted or broad-spectrum binders, and isolate binding sites by profiling single domains, truncated, or full-length forms of the target protein.[37] Differential binding modalities may in turn provide distinct degrader-induced target presentation for ubiquitylation.[38] Although traditional affinity-based screening methodologies support parallel and binding site agnostic approaches, the vast scale of DELs, containing millions to trillions of ligands, facilitates rapid exploration of large and diverse chemical space that otherwise remains inaccessible to anything other than virtual screening.[39–42]

The development of bifunctional degraders provides an ideal venue for affinity-based DEL screening as both a POI and E3 ligases can be screened in parallel to identify POI or E3 ligase selective binders (Figure 7.1A). Once binding is confirmed, a bifunctional degrader is synthesized by connecting the two ligands by a chemical linker.[43] Ligands identified *via* traditional high-throughput screening (HTS) and fragment-based screening methods require additional investigation including structure determination or structure–activity exploration to identify tolerated linker attachment points,

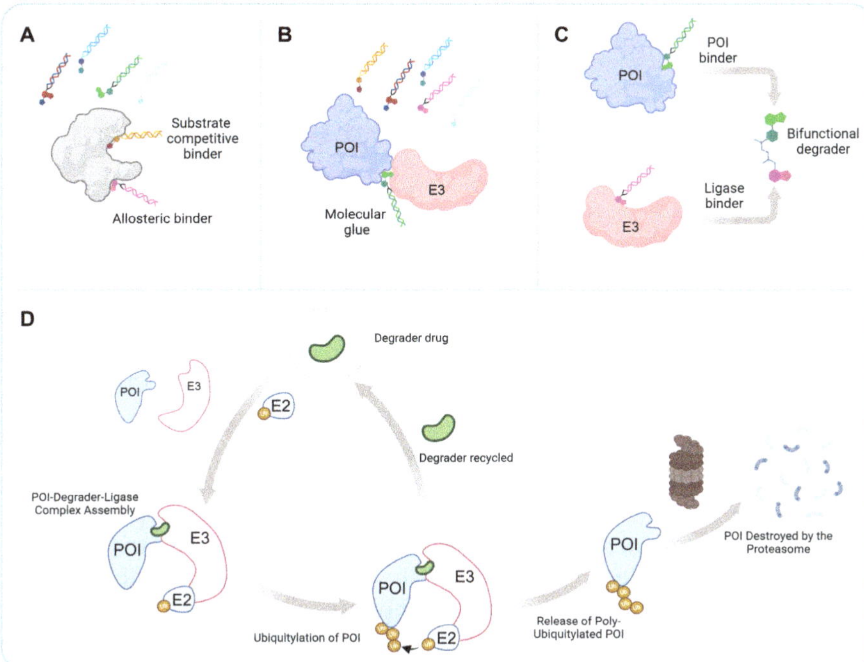

Figure 7.1 (A) DNA-encoded library screens probe and identify binders to multiple binding pockets. These binders may be inhibitory, substrate or degron competitive, or allosteric binders, all of which have the potential for targeted protein degradation applications. (B) DNA-encoded library screens can be leveraged to identify molecular glues, which bind cooperatively to a PPI interface of a POI and E3 ubiquitin ligase either by stabilizing a native low affinity interaction or by inducing a neomorphic interaction. (C) DNA-encoded library discovery can be applied to the discovery of both POI and E3 ubiquitin ligase binders, which are then linked together to generate a bifunctional degrader. (D) Targeted protein degradation functions through the association of a POI with an E3 ubiquitin ligase, induced by a small molecule. This association, in conjunction with an E2 ubiquitin ligase, leads to the ubiquitylation of the POI, marking it for degradation by the proteasome.[48] Created with Biorender.com.

a process that can add significant time and cost to degrader development. By nature of their design, DEL molecules contain a solvent exposed, synthetically accessible DNA attachment point, immediately highlighting a starting vector for bifunctional degrader linker installation.[44] Targeted screening of molecular glues for a specific POI and E3 ligases requires the inclusion of both POI and E3 ligase components during screening (Figure 7.1B), achievable through modifications to the typical affinity based DEL screening protocol or alternative one-bead-one-compound DEL strategies in which the DNA-encoded compound is released from the bead and DNA-tag for solution phase activity based screening.[45,46] Serendipitous discovery of molecular glues through POI or E3 ligase screening and subsequent downstream characterization have also been documented.[47]

7.3 DELs for the Identification of E3 Ubiquitin Ligase Binders

Additional E3 ubiquitin ligase binder identification is essential for advancing the current opportunities for bifunctional degrader discovery and optimization. Despite the predominance of CRBN- and VHL-based degraders in the field, there is a clear need for novel ligase binders to expand target accessibility, enhance degrader selectivity in terms of tissue specific or tissue sparing ligase profiles, and potentially mitigate CRBN- or VHL-based resistance.[49,50] E3 ligase binders have additional applications as inhibitors. Inhibiting key ligase–substrate interactions can increase natural substrate concentrations by preventing protein turnover. This mechanism serves as a strategic approach for modulating metabolic and cancer pathways.[51–56] Analogous to DEL screening of enzymes where potential inhibitors are identified *via* screening of apo protein in parallel with an inhibitor bound-target condition to identify inhibitor competitive ligands, E3 ligase substrate degrons (the minimal elements of a protein that are required for binding to the E3 ligase) can be used as DEL screen additives to identify binders to degron interacting sites.[57–59] Substrate competitive binders have the potential for ligase inhibitor development, while both competitive and non-competitive binders have the potential for degrader development, provided they do not inhibit ligase function. In this approach, the E3 ligase, in its apo form, is screened alongside the same ligase in the presence of a saturating concentration of a known binder, substrate peptide, or interacting partner. DEL ligands enriched in apo protein conditions, but not in the presence of the known interactor, are likely competitive binding across the degron binding surface. Degron competitive binders can then be confirmed using biochemical FRET or FP displacement assays of the labelled known E3 ligase degron peptide.

This strategy is exemplified by DEL screening of glucose-induced degradation protein 4 homolog (GID4), a substrate receptor of the CTLH E3 ligase complex, using the 4.4 billion member DELopen library.[58] Wild-type GID4 was initially screened with the PGLWKS degron peptide, but the low affinity binding of the PGLWK2 degron peptide ($K_D = 4$ μM by FP) led to incomplete saturation of the substrate binding site. An alternative approach leveraging DEL screening of the $GID4^{Q282A}$ mutant, which does not bind the PGLWKS peptide, identified a subset of binders with 2-fold lower enrichment compared to the wild type. These hits were resynthesized and validated for GID4 binding *via* thermal shift and PGLWKS-peptide FP competition assays (IC_{50} = 5.4, 4.8 and 4.7 μM). Further analysis by isothermal titration calorimetry (ITC) confirmed binding to wild-type GID4 (K_D of 5.6 μM) and enhanced affinity of the GID4 mutant (K_D of 0.28 μM). Cocrystallization and X-ray crystallography (PDB: 7U3J, 7U3K, 7U3L) confirmed the competitive binding mode of the degrons. This work establishes a foundation for the utilization of the GID4 CTLH complex in developing degraders.

DCAF1-based PROTACs have demonstrated sustained degradation of clinically relevant targets such as BRD9 and BTK, even in cell lines that have developed resistance to VHL or CRBN.[3,49] The DDB1-CUL4-associated factor (DCAF1), a substrate receptor in the CUL4A-DDB1-DCAF1 ($CRL4^{DCAF1}$) complex, is nuclear localized; both DCAF1 and DDB1 show similar high tumor essentiality, which is an ideal E3 ligase profile for targeted protein degradation in cancer.[60] DCAF1 is hijacked by the viral human immunodeficiency virus 1 (HIV1) protein, Vpr, inducing cell cycle arrest. This protein functions in substrate recognition as part of the EED-DDB1-VPRBp (EDVP) complex and is shown to phosphorylate histone H2A.[61] Although the DCAF1-based bifunctional degraders highlighted by Schröder *et al.* were identified from focused iterative screening and optimization of small libraries, a simultaneous independent DEL-driven hit identification campaign identified binders to the same pocket that could easily be translated to bifunctional degraders.[59,62]

DEL screening of DCAF1 with a 114 billion DNA encoded library focused on the wild type DCAF1 WDR domain, leveraging machine learning guided hit selection from vendor catalogs and structure guided optimization to identify OICR-8269, a nM binder of DCAF1.[59,63] Initial confirmation by differential scanning fluorimetry (DSF) and confirmation by surface plasmon resonance (SPR) identified the vendor catalog hit Z139123226, a 12 ± 4 μM binder. Investigation of the preliminary hit by X-ray crystallography highlighted suboptimal chiral geometry. Following molecular modelling studies, additional enantiomers were tested, yielding enantiomer 3d, a 490 nM binder. Compound 3d binds deep in the central tunnel of the DCAF1 WDR domain ring. Comparison with the published crystal structure of DCAF1 in complex with the viral protein Vpx (PDB ID: 5AJA) indicates binding at a site distant from the Vpx peptide binding site. Subsequent molecular modeling-based optimization yielded the optimized hit OICR-8269, which binds DCAF1 with a K_D of 38 ± 2 nM, as measured by SPR. Cellular thermal shift assay (CETSA)-based target engagement using the NanoLuc split luciferase technology (Promega) confirmed dose-dependent stabilization of DCAF1-WDR but not full length DCAF1 by OICR-8269. This work represents part of a broader study applying machine learning guided DEL hit selection followed by high throughput crystallography for identification of binders from the broader WDR protein family, a strategy that has the potential for highly parallelized ligase binder identification.[64]

The use of irreversible covalent ligands in bifunctional degraders (for both the POI and ligase) demonstrated advantages over reversible binders in some situations. These include the ability to identify ligands to targets and ligases otherwise intractable to screens that utilize traditional compound collections, enhanced selectivity, and improved kinetics of ternary complex formation and degradation.[65,66] Covalent hit ID through DEL screening is a well-established and broadly applied aspect of the technology, and covalent DEL hits confer the same benefits in terms of providing solvent exposed handles for the synthesis of hetero-bifunctional molecules.[67–70]

7.4 DELs for the Identification of Target Binders for TPD

Identification of high affinity target binders and inhibitors across diverse target classes using DNA-encoded library screening has been well exemplified.[36,41,71–77] Unlike traditional inhibitor screens, DEL screening for degrader development is agnostic of the binding site and can equally leverage active site or allosteric binders to achieve degradation.[38,78,79] Conversion of DEL target binders to bifunctional degraders for profiling through installation of a linker and a ligase binder is rapid and can leverage the known vector and linker accessible chemistry of the original DEL construction (Figure 7.1C).[44] Bifunctional degrader optimization then fine tunes the properties of the degrader as a whole to achieve optimal ternary complex formation, ubiquitylation, solubility, permeability, and oral bioavailability, all while maintaining binding affinity to both the target and the ligase.[12,80] Although final bifunctional degrader optimization can fine-tune these properties, the physiochemical properties of the initial E3 ligase and target binding moieties will dictate the overall degrader profile and potential for oral bioavailability during late stage optimization, highlighting the need for property conscious ligand identification at the screening stage.[12]

The utility of DEL screening in bifunctional degrader development was exemplified by Disch *et al.* in the DEL-based discovery of new estrogen receptor α (ERα) binders, which induce degradation of ERα and three gain of function point mutants when converted to bivalent degraders.[44] The oral, CRBN-based ERα bivalent degrader Vepdegestrant (ARV-471) was the first bivalent degrader to enter phase III clinical trials and has shown promising anti-tumor activity in ER+/HER2− breast cancer, cementing the interest in this target for degrader development.[14] More broadly, ERα mutations play a key role in ERα inhibitor-based therapeutic resistance in breast cancer. Degraders that bind ERα agnostic of the mutation state may provide alternative therapies for next-generation breast cancer treatment. Disch *et al.* leverage the parallelized nature of DEL screening to investigate WT ERα and three gain of function mutants S463P, Y537S, and D538G. Binders to all target forms were identified, including a chlorophenol hit series with IC_{50} values ranging from 7.2 to 110 nM against WT and all three mutants, which demonstrated modest cellular agonist activity. Leveraging the known DNA attachment point and docking-based design to inform linker placement, the authors then profiled a set of alkyl, methoxy or methoxy phenyl linkers for ERα binding affinity and cellular activity. A methoxy phenyl linker vectored off the DNA attachment point was identified as the ideal vector location. Alternate linkers and attachment points resulted in either lower affinity for

ERα *in vitro* or agonist activity of ERα *in vivo*. Rapid synthesis and profiling of degradation was achieved by leveraging CRBN, IAP and VHL binders with alkyne functionalized 1–3 PEG unit linkers linked to an azide-functionalized DEL or literature-derived ERα binders *via* Huisgen 1,3-dipolar cycloaddition.[81] Potent degradation of ERα with nM DC_{50} values was achieved using a VHL-based degrader linked to the DEL derived ERα binder in ER-positive cells. This work demonstrates the rapid translation of DEL hits to bifunctional degraders by employing rapid, modular "click" chemistry linkers using DEL headpiece attachment vectors for linker installation.

DEL-screening informed bifunctional degrader construction was also exemplified in the development of a receptor interacting protein kinase 2 (RIPK2) degrader.[82] GSK leveraged a combination of high throughput screening (HTS) and DEL to identify potent and selective inhibitors of RIPK2, including a key shared pharmacophore between HTS and DEL series.[83] Structure–activity relationship (SAR) data from historical kinase DEL screens were leveraged to identify modifications to the HTS hit that improved selectivity.[84] The DNA attachment point of the DEL hit informed linker attachment for the development of a potent and selective VHL-based RIPK2 PROTAC.[85] Additional IAP- and CRBN-based PROTACs were profiled with similar potencies to the VHL PROTACs.[80] The IAP PROTACs were optimized to produce degraders with extended PD responses. When formulated into a slow-release matrix, the lead IAP PROTAC maintained >90% degradation of RIPK2 *in vivo* for >40 days.[86]

7.5 DEL Screening for Degraders and Glues

Modifications to traditional DEL screening have adapted the technology to accelerate bifunctional degrader discovery by directly identifying compounds that form productive ternary complexes, bypassing time-consuming linker and POI-binder optimization. A recent proof-of-concept study from the Schrieber Group aiming to identify a BRD4 degrader uses this strategy, which they term "CIP-DEL" (Chemical Inducer of Proximity).[19] Using the established BRD4 degrader MZ1 as a template for the library design, the DNA headpiece is attached at a solvent exposed vector identified from previously published MZ1 ternary complex crystal structures (PDB: 5T35).[87] The VHL-targeting portion of the established BRD4 degrader MZ1 is functionalized with a set of linkers with varied lengths and compositions. These VHL ligand-connectors are encoded by unique DNA barcodes (cycle 1) and are functionalized with 2-cycle BRD4 binders through split-and-pool synthesis. The final one-million member CIP library is screened against BRD4 with (+) and without (−) VHL-elongin C-elongin B (VCB) to identify compounds that bind BRD4 in the presence of VCB. The hypothesis is that BRD4 binders identified in the −VCB condition would have reduced enrichment in the +VCB condition if the compound could not induce a stable ternary complex (Figure 7.2A). Analysis of the screen output shows relatively even barcode count distribution for each linker in the −VCB condition, compared to a

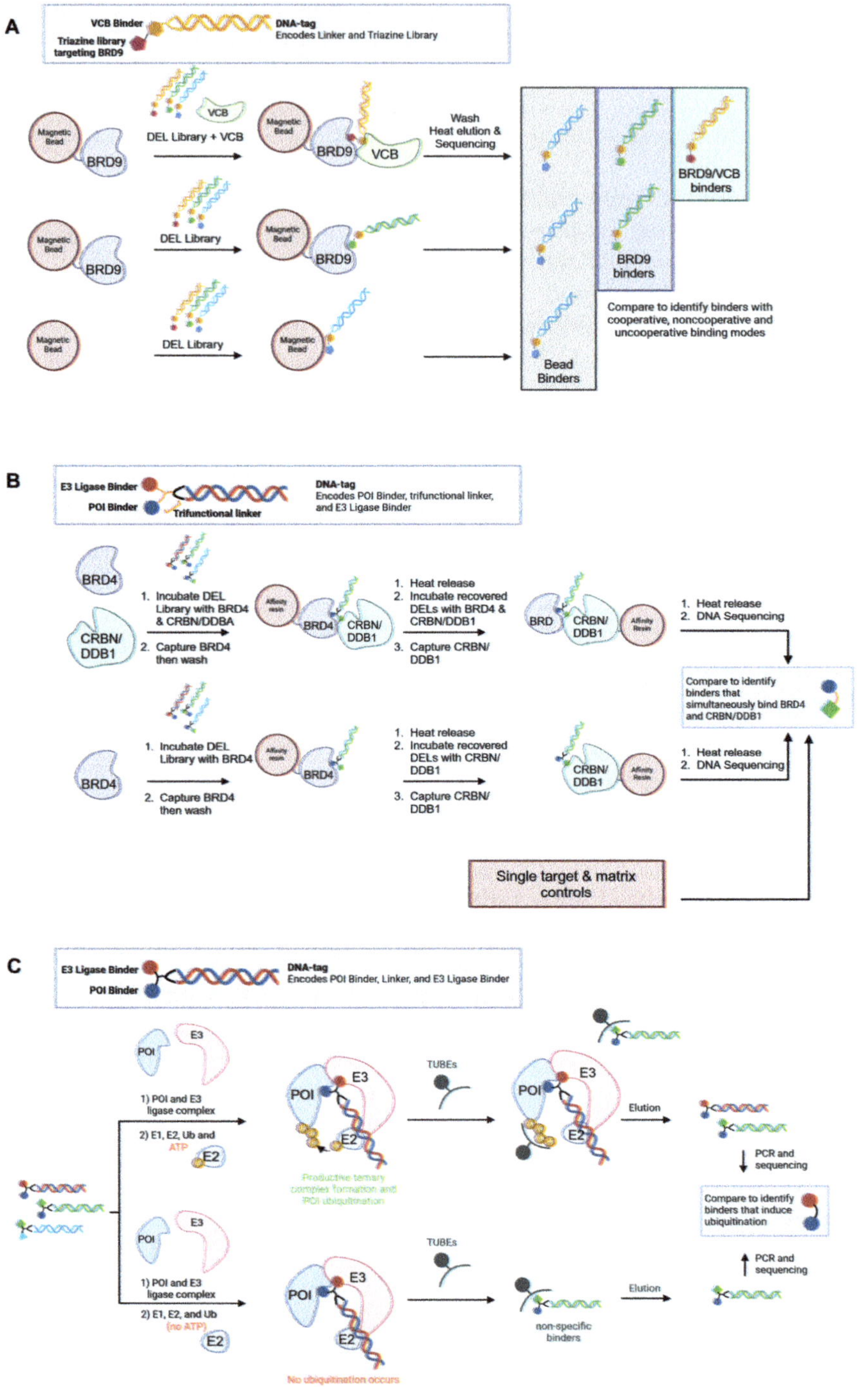
A
VCB Binder
Triazine library targeting BRD9
DNA-tag
Encodes Linker and Triazine Library
Magnetic Bead
BRD9
VCB
DEL Library + VCB
Wash
Heat elution &
Sequencing
DEL Library
BRD9/VCB binders
BRD9 binders
Bead Binders
Compare to identify binders with cooperative, noncooperative and uncooperative binding modes
B
E3 Ligase Binder
POI Binder
Trifunctional linker
DNA-tag
Encodes POI Binder, trifunctional linker, and E3 Ligase Binder
BRD4
CRBN/DDB1
1. Incubate DEL Library with BRD4 & CRBN/DDBA
2. Capture BRD4 then wash
Affinity resin
1. Heat release
2. Incubate recovered DELs with BRD4 & CRBN/DDB1
3. Capture CRBN/DDB1
BRD
Affinity Resin
1. Heat release
2. DNA Sequencing
Compare to identify binders that simultaneously bind BRD4 and CRBN/DDB1
1. Incubate DEL Library with BRD4
2. Capture BRD4 then wash
1. Heat release
2. Incubate recovered DELs with CRBN/DDB1
3. Capture CRBN/DDB1
Single target & matrix controls
C
E3 Ligase Binder
POI Binder
DNA-tag
Encodes POI Binder, Linker, and E3 Ligase Binder
POI
E3
1) POI and E3 ligase complex
2) E1, E2, Ub and ATP
E2
Productive ternary complex formation and POI ubiquitination
TUBEs
Elution
PCR and sequencing
Compare to identify binders that induce ubiquitination
2) E1, E2, and Ub (no ATP)
No ubiquitination occurs
non-specific binders

dramatic shift towards compounds with a C15 linker in the +VCB condition, suggesting the C15 linker is critical for stable ternary complex formation. Overall, a potent degrader with a DC_{50} of 17 nM was identified directly from the screen without any further optimization.

HitGen employs a similar strategy in their design of a "PROTAC DEL" used to identify BRD4 degraders.[35] All compounds in this PROTAC DEL contain thalidomide-based CRBN binders as the E3 ligase component, combined with various trifunctional linkers and BRD4 binders that are encoded by the DNA barcode. The HitGen approach focuses on screening of BRD4 and CRBN in tandem using orthogonal affinity tags, with BRD4 immobilized in one round and CRBN immobilized in the second round (Figure 7.2B). In this manner, PROTACs that bind to both BRD4 and CRBN are identified, and the final DEL signal directly reflects the ternary complex binding cooperativity of the PROTAC with the POI and E3 ligase. Cpd1, identified through this screen, displays proteasome-dependent BRD4 degradation in the MV4-11 cell line and shows stronger activity than dBET1 in apoptosis and anti-proliferation assays, demonstrating the feasibility of dual-protein affinity selection with PROTAC DEL for PROTAC discovery.

The Schrieber Group's CIP-DEL strategy is equally applicable for molecular glue degrader screening, exemplified through a screen of BRD9 in +VCB and −VCB conditions.[89] A major advantage of molecular glue degraders over conventional bifunctional degraders is the lack of a "hook effect."[91] At high degrader concentrations, bifunctional degraders will form binary complexes with either the POI or the E3, competing with formation of the ternary complex, because the POI binder and E3 binder components have a higher affinity for their respective ligands compared to the degrader's

Figure 7.2 Strategies for screening of bifunctional degrader DELs. (A) Cooperativity-based DEL screening of BRD4 with and without VCB. BRD4 is first immobilized and then incubated with DEL and VCB. The output is then compared to samples containing BRD4 alone, or magnetic resin alone, to identify cooperative, noncooperative and uncooperative DEL members that bind BRD4 and VCB.[88,89] (B) Tandem screening of a thalidomide PROTAC DEL using orthogonally tagged BRD4 and CRBN. The library is incubated with BRD4 and CRBN/DDB1, then captured *via* BRD4 immobilization.[35] Following heat elution, the library is incubated once more with BRD4 and CRBN/DDB1, then captured *via* DDB1. The output is heat eluted. The sequencing output is compared to a sample in which the library is first incubated with BRD4, the CRBN/DDB1 sequentially, as well as matrix and target only outputs, to identify library members that bind to BRD4 and CRBN/DDB1 simultaneously. (C) Ubiquitylation-based DEL screening. PROTAC DEL is incubated with POI and the E3 ligase complex, followed by addition of E1, E2, Ub and ATP.[90] Following transfer of ubiquitin to the POI, ubiquitylation competent complexes are captured using TUBEs for sequencing. The output is compared to the library incubated with POI, the E3 ligase complex, E1, E2, and Ub without ATP followed by TUBE pulldown to identify non-specific binders.[48] Created with Biorender.com.

affinity for the complex. This results in the observed "hook effect", where POI degradation is reduced at high degrader concentrations. Conversely, molecular glues bind cooperatively to both the POI and E3; the affinity of the glue for the ternary complex is higher than its affinity to either the POI or E3 alone, with no defined hook effect observed. In this BRD9 screen, the "presenter ratio," or the ratio of ternary enrichment (enrichment of a compound in the presence of the target and E3) to binary enrichment (enrichment of a compound in the presence of the target only), is used as a predictive measure of cooperativity. A higher presenter ratio suggests a compound with higher cooperativity, as binding to BRD9 is increased in the presence of VCB. Compounds identified through the screen with high presenter ratios, most notably within the 13 series, show no significant hook effect in AlphaScreen assays, demonstrating cooperative binding. Within this series, 13-7 displays a cooperativity factor similar to rapamycin, a classical molecular glue. 13-7 was then demonstrated to induce dose-dependent formation of ternary complexes in cells (EC_{50} of 17 nM) and degrades BRD9–HiBiT fusion proteins in HEK293T cells (DC_{50} of 108 nM) without any significant hook effect.

Conventional DEL screens have also yielded serendipitous discovery of molecular glues. Following a library screen of 65 billion DELs against the ATAD2 bromodomain (BD), hit-to-lead optimization identified BAY-850, a potent inhibitor that induces ATAD2 BD dimerization.[47] BAY-850's binding mode was examined using non-denaturing mass spectrometry revealing that the compound induces ATAD2 dimer formation in a concentration dependent manner, with no observed binding to monomeric ATAD2. No binding to either monomeric or dimeric ATAD2 is observed using the inactive control BAY-460. A BROMOscan panel follow-up indicates that BAY-850 exclusively hits ATAD2 and notably not ATAD2B, demonstrating strong selectivity over closely related targets. In addition to new strategies designed for rational screening of molecular glues, this example highlights the utility of conventional affinity-based DEL selection in the discovery of potent, isoform selective molecular glues.

Recent advances in activity-based DEL screening technologies present opportunities for DEL screens for molecular glues and bifunctional degraders.[45] Conference presentations from Plexium highlighted the use of their DELPhe platform to identify monovalent glue degraders of BRD4 through one-bead-one-compound (OBOC) DEL screening in pico-wells. Using this strategy, DEL members are released from beads, while the DNA-tag remains resin bound, and interrogated in cell-based BRD4-reporter assays in pico-wells, monitoring for BRD4 degradation. Library beads from pico-wells with active degraders are then sequenced to identify the compound structures. In contrast to solution-based DEL screening for molecular glues, this strategy de-couples the chemical library member from the DNA-encoding tag, removing the potential steric liability of the DNA-encoding tag, which may occlude complex formation.

Two recent pre-prints from Huang *et al.* and Cai *et al.* leverage solution phase, activity-based DEL screening for binders that induce *in vitro* ubiquitylation.[90,92] Using this strategy, a solution phase monovalent or bivalent degrader DEL is incubated with the POI and E3 ligase complex, and then E1, E2, Ub, and ATP are supplemented to allow for *in vitro* ubiquitylation (Figure 7.2C). Unlike traditional, affinity-based DEL selections in which the POI is captured, here ubiquitylated ternary complexes are captured with tandem ubiquitin binding entities (TUBEs), specifically isolating compounds that generate ternary complexes productive for ubiquitylation. These advances provide an avenue for augmenting affinity-based DEL selections with downstream activity profiling to drive degrader optimization.

In a study by Huang *et al.*, the authors first verify that DNA-conjugated PROTACs (composed of the BRD4 inhibitor JQ1, the VHL ligand VH032, and a trifunctional linker) can induce POI ubiquitylation and degradation in a model system with the BD1 domain of BRD4 as the POI and VHL as the E3 ligase.[90] The control degrader PTC-5 induces significant BRD4 BD1 ubiquitylation when tested in an *in vitro* ubiquitylation assay. When transfected into HEK293T cells, the most potent compound PTC-5# exhibits a DC_{50} of ~4 nM and a D_{max} of 99% after 24 h. After successfully confirming POI ubiquitylation and degradation, a PROTAC-DEL of 950 "E3L-linker-POI-L" combinations is screened against BRD4 BD1, with (+) and without (−) ATP, with the −ATP condition serving as a negative control for ATP-dependent ubiquitylation. From this screen, 1-1-1 and 15-1-3 show significant enrichment in the +ATP condition compared to the −ATP condition; in HEK293T cells, 1-1-1 exhibits a DC_{50} of 9.7 nM and a D_{max} of 99%, and 15-1-3 exhibits a DC_{50} of 60 nM and a D_{max} of 66%. Furthermore, they identify 4-4-10 as an isoform-selective degrader for short BRD4 with a DC_{50} of 0.26 μM. The library was orthogonally screened using a conventional nickel affinity pull-down of the His-tagged BD1, hypothesized to enrich compounds that form stable ternary complexes, which may or may not lead to ubiquitylation. They observe that both active and inactive degraders are enriched, suggesting that screening for stable ternary complexes is not a reliable predictor of protein ubiquitylation and degradation, consistent with other observations.[93,94]

In a study by Cai *et al.*, a single-stranded DNA-library is first hybridized with a DNA-labelled POI, followed by addition of a reconstituted E3 ligase system.[92] This single stranded dual POI and DEL encoding strategy induces proximity between the DEL and POI, with applications for [library] × [library] monovalent or bivalent degrader screening strategies. Library members that induce productive ubiquitin transfer are then captured using an anti-ubiquitin affinity-based enrichment strategy. Using this strategy, the authors demonstrate selective ubiquitin transfer and enrichment of four IKZF2b-CRBN molecular glue–DNA conjugates over a control amine–DNA conjugate, further observing that the most potent molecular glue tested, NVP-DKY709, demonstrated the highest enrichment.[95]

7.6 Discussion and Perspectives

DEL screening provides an optimal tool for bifunctional degrader discovery through rapid selection assays for both ligase and target binder identification. The binding site agnostic nature of DEL screening is ideal for target or ligase binder identification as degradation is not constrained to active site binders. Multiple binding sites and series offer numerous advantages for degrader generation including differential ternary complex formation and lysine presentation for ubiquitylation. In the absence of structure-based design, the DEL DNA-attachment point provides a preliminary solvent exposed vector for degrader synthesis enabling rapid degrader matrix generation.[44] Hit selection of target or ligase binders from lower molecular weight and more ligand efficient libraries may be beneficial for bifunctional degrader generation as the ligase–linker–target combination may still require extensive optimization, which is limited by the physiochemical properties of the initial binders to reach the property space required for oral bioavailability.[12,96] Identification of covalent fragment ligase binders has recently shown promise as a means for lower MW starting points for ligase binder optimization and early degradation confirmation, a strategy that leverages the catalytic nature of ligase target ubiquitylation by reducing the off-rate to increase ligase residence time and translates readily to covalent DEL screen applications covered elsewhere in this book.[67–70,97–99] Despite this challenge, new techniques in cell-based screening assays, rapid automated synthesis, *in vitro* profiling, and compound optimization are well suited to the modularity of bifunctional degrader synthesis and optimization.[100–102]

DEL screening of degrader libraries for productive ternary complex formation or target ubiquitylation provides an opportunity for DEL-based ligase–linker–target binder optimization using focused libraries or molecular glue identification from unbiased DELs. Focused libraries are generally geared towards a single target and ligase; however, linker–ligase or linker–target sub-libraries can be easily modified as "on-demand" libraries for new targets or ligases. Parallelized DEL screening of diverse ligase-binder containing bifunctional DEL degrader libraries could be leveraged to rapidly profile many interacting partners from a single library. Although ternary complex formation screens geared towards cooperativity can provide valuable information on linker length and the attachment vector, ternary complex formation does not directly translate to degradation.[93,94] Recent advances in *in vitro* ubiquitylation activity-based DEL screening bridge the gap between ternary complex formation and productive ubiquitylation and are applicable to both focused bifunctional degrader library screening and unbiased molecular glue screening; however, this approach may be technically limited for ligases with more complex ubiquitylation machinery.[90] Although both ternary complex formation and ubiquitylation based activity screening provide initial lead bifunctional or molecular glue degraders, the DNA-attachment point intrinsic to DEL-based discovery may cause steric hindrance preventing ternary complex formation and resulting in false negatives and loss of valuable structure–activity relationships. The crystal structures of bifunctional degrader and molecular glue driven ternary complexes of CDK12:DDB1, and calcineurin complexed with cyclosporin A and cyclophilin show limited solvent accessibility, suggesting DEL-based screening of ternary complexes and ubiquitylation may not work for every target system.[103–105] Identification of ideal systems for screening could benefit from computational approaches to predict ternary complex complementarity or empirical, at-risk, multiplexed screening of ligase and target combinations.[26,106]

Identification of ligase and target ligands using DEL screening methodologies provides ready access to chemical inputs suited for degrader synthesis.[100–102] Combinatorial bifunctional degrader synthesis workflows can systematically profile target and ligase binders spanning multiple binding sites and linker vectors, enabling rapid and broad sampling of combinations to optimize ternary complex formation, ubiquitylation, and degradation. Although the DNA attachment point provides a preliminary solvent exposed vector for degrader synthesis, SAR exploration can expose new vector options within a chemical series to be explored in degrader optimization. Downstream hit-to-lead degrader optimization relies on simultaneous ligase–linker–target binder optimization, and many optimized degraders utilize very small highly conformationally constrained linkers.[82,107] Exemplified DNA-encoded combinatorial strategies for bifunctional and glue discovery and optimization are constrained by the chemical linkage to the encoding DNA, which limits the linker vector and ternary complex orientations that can be explored, potentially missing the most optimal configurations.[35,45,88,89,92] One-bead-one-compound DELs provide an alternative encoding method for combinatorial degrader optimization, where the library members are released with minimal scarring and without the same steric liabilities as a DNA conjugated linker.[45,46]

Current exemplified strategies leveraging DELs for ligase binder discovery for degrader development focus on clearly defined ligase and target combinations.[35,58,59,88–90] Future development in this field could leverage existing DEL-derived ligase binders and SARs to identify and profile novel ligase–target molecular glue interaction, neo-substrate interactions, and potential liabilities.[92,103,108,109] Recent work investigating a diverse library of thalidomide analogs in patient-derived leukemia and medulloblastoma cell lines to identify novel targets and CRBN modulators with potential liabilities demonstrates that structural modifications of known ligase binders or molecular glues can change the profile of neo-substrates. A similar strategy was employed by the Winter lab to derivatize and optimize known CDK12 binders for CDK12:DDB1 glue activity leading to cyclin K degradation.[103] Such a strategy is complementary to a DEL-based discovery approach where a diverse SAR is represented in the DEL itself, and a secondary neo-substrate profiling screen following ligase binder or molecular glue identification has the potential to identify novel target repertoires or flag undesired degradation targets early in the discovery process.

DNA-encoded libraries provide a vast source of chemical diversity ideal for novel ligase and target binder discovery and degrader development. New advances in DEL screening strategies provide a foundation for ternary complex screening and ubiquitylation activity-based readout of bifunctional degrader libraries that can be adapted on demand for each ligase and target binder system leveraging the modular combinatorial nature of DEL construction. Degrader discovery presents an ideal challenge to showcase and advance DEL construction and screening strategies to provide layered SAR data beyond initial binder/degrader discovery to rapidly advance hit-to-lead optimization.

TPD is part of a growing field of proximity-inducing modalities, many of which have emerged over the last few years.[110–112] These induced proximity approaches open avenues to previously unexplored pharmacology. More recent TPD strategies leveraging non-E3 ligase mediated degradation pathways include lysosomal recruitment (LYTACs), the autophagy lysosomal system (AUTOTACs), and direct proteosome recruitment (ByeTACs).[113–117] The qualities of DEL screening described here that make it an ideal fit for TPD applications also make this technology well-suited to other induced proximity modalities.

References

1. K. M. Sakamoto, K. B. Kim, A. Kumagai, F. Mercurio, C. M. Crews and R. J. Deshaies, *Proc. Natl. Acad. Sci. U. S. A.*, 2001, **98**, 8554–8559.
2. M. Békés, D. R. Langley and C. M. Crews, *Nat. Rev. Drug Discovery*, 2022, **21**, 181–200.
3. S. Montoya, J. Bourcier, M. Noviski, H. Lu, M. C. Thompson, A. Chirino, J. Jahn, A. K. Sondhi, S. Gajewski, Y. S. M. Tan, S. Yung, A. Urban, E. Wang, C. Han, X. Mi, W. J. Kim, Q. Sievers, P. Auger, H. Bousquet, N. Brathaban, B. Bravo, M. Gessner, C. Guiducci, J. N. Iuliano, T. Kane, R. Mukerji, P. J. Reddy, J. Powers, M. Sanchez Garcia de Los Rios, J. Ye, C. Barrientos Risso, D. Tsai, G. Pardo, R. Q. Notti, A. Pardo, M. Affer, V. Nawaratne, T. M. Totiger, C. Pena-Velasquez, J. M. Rhodes, A. D. Zelenetz, A. Alencar, L. E. Roeker, S. Mehta, R. Garippa, A. Linley, R. K. Soni, S. S. Skanland, R. J. Brown, A. R. Mato, G. M. Hansen, O. Abdel-Wahab and J. Taylor, *Science*, 2024, **383**, eadi5798.
4. C. Kim, X. D. Wang, Z. Liu, S. Zha and Y. Yu, *Biochemistry*, 2023, **62**, 561–563.
5. L. Fu, J. Zhang, B. Shen, L. Kong, Y. Liu, W. Tu, W. Wang, X. Cai, X. Wang, N. Cheng, M. Xia, T. Zhou, Q. Liu, Y. Xu, J. Yang, P. Gavine, U. Philippar, R. Attar, J. P. Edwards, J. D. Venable and X. Dai, *J. Med. Chem.*, 2021, **64**, 10878–10889.
6. S. Wang, L. Han, J. Han, P. Li, Q. Ding, Q. J. Zhang, Z. P. Liu, C. Chen and Y. Yu, *Nat. Chem. Biol.*, 2019, **15**, 1223–1231.
7. D. Menolfi and S. Zha, *Cell Biosci.*, 2020, **10**, 8.
8. B. Adhikari, J. Bozilovic, M. Diebold, J. D. Schwarz, J. Hofstetter, M. Schroder, M. Wanior, A. Narain, M. Vogt, N. Dudvarski Stankovic, A. Baluapuri, L. Schonemann, L. Eing, P. Bhandare, B. Kuster, A. Schlosser, S. Heinzlmeir, C. Sotriffer, S. Knapp and E. Wolf, *Nat. Chem. Biol.*, 2020, **16**, 1179–1188.

9. X. Liu, A. F. Kalogeropulou, S. Domingos, N. Makukhin, R. S. Nirujogi, F. Singh, N. Shpiro, A. Saalfrank, E. Sammler, I. G. Ganley, R. Moreira, D. R. Alessi and A. Ciulli, *J. Am. Chem. Soc.*, 2022, **144**, 16930–16952.
10. T. Tashima, *Antibodies*, 2023, **12**, 43.
11. C. Kofink, N. Trainor, B. Mair, S. Wohrle, M. Wurm, N. Mischerikow, M. J. Roy, G. Bader, P. Greb, G. Garavel, E. Diers, R. McLennan, C. Whitworth, V. Vetma, K. Rumpel, M. Scharnweber, J. E. Fuchs, T. Gerstberger, Y. Cui, G. Gremel, P. Chetta, S. Hopf, N. Budano, J. Rinnenthal, G. Gmaschitz, M. Mayer, M. Koegl, A. Ciulli, H. Weinstabl and W. Farnaby, *Nat. Commun.*, 2022, **13**, 5969.
12. K. R. Hornberger and E. M. V. Araujo, *J. Med. Chem.*, 2023, **66**, 8281–8287.
13. C. A. Lipinski, F. Lombardo, B. W. Dominy and P. J. Feeney, *Adv. Drug Delivery Rev.*, 2001, **46**, 3–26.
14. Arvinas, San Antonio Breast Cancer Symposium, 2022.
15. A Study to Learn About a New Medicine Called Vepdegestrant (ARV-471, PF-07850327) in People Who Have Advanced Metastatic Breast Cancer (VERITAC-2): https://www.clinicaltrials.gov/study/NCT05654623, (accessed April 2024).
16. Z. Hu and C. M. Crews, *ChemBioChem*, 2022, **23**, e202100270.
17. X. Gao, H. A. Burris, J. Vuky, R. Dreicer, A. O. Sartor, C. N. Sternberg, I. J. Percent, M. H. A. Hussain, A. R. Kalebasty, J. Shen, E. I. Heath, G. Abesada-Terk, S. G. Gandhi, M. McKean, H. Lu, E. Bernghorm, R. Gedrich, S. D. Chirnomas, N. J. Vogelzang and D. P. Petrylak, ASCO, 2022.
18. L. Ackerman, G. Acloque, S. Bacchelli, H. Schwartz, B. J. Feinstein, P. La Stella, A. Alavi, A. Gollerkeri, J. Davis, V. Campbell, A. McDonald, S. Agarwal, R. Karnik, K. Shi, A. Mishkin, J. Culbertson, C. Klaus, B. Enerson, V. Massa, E. Kuhn, K. Sharma, E. Keaney, R. Barnes, D. Chen, X. Zheng, H. Rong, V. Sabesan, C. Ho, N. Mainolfi, A. Slavin and J. A. Gollob, *Nat. Med.*, 2023, **29**, 3127–3136.
19. A. Starodub, A. Gollerkeri, C. D. savi, J. Dey, S. Agarwal, S. Donohue, R. Perea, C. Klaus and J. Gollob, *J. Clin. Oncol.*, 2022, **40**, TPS3171.
20. D. W. Robbins, M. A. Noviski, Y. S. Tan, Z. A. Konst, A. Kelly, P. Auger, N. Brathaban, R. Cass, M. L. Chan, G. Cherala, M. C. Clifton, S. Gajewski, T. G. Ingallinera, D. Karr, D. Kato, J. Ma, J. McKinnell, J. McIntosh, J. Mihalic, B. Murphy, J. R. Panga, G. Peng, J. Powers, L. Perez, R. Rountree, A. Tenn-McClellan, A. T. Sands, D. R. Weiss, J. Wu, J. Ye, C. Guiducci, G. Hansen and F. Cohen, *J. Med. Chem.*, 2024, **67**, 2321–2336.
21. E. Searle, F. Forconi, K. Linton, A. Danilov, P. McKay, D. Lewis, D. El-Sharkawi, M. Gleeson, J. Riches, S. G. Injac, R. Shih, S. Nandakumar, M. Tan, G. Cherala, E. Meredith and G. P. Collins, ASH, 2023.
22. Arvinas Enters into a Transaction with Novartis, including a Global License Agreement for the Development and Commercialization of PROTAC® Androgen Receptor (AR) Protein Degrader ARV-766 for the

Treatment of Prostate Cancer: https://ir.arvinas.com/news-releases/news-release-details/arvinas-enters-transaction-novartis-including-global-license, (accessed June 2024).
23. M. McKean, A. I. Spira, E. Rosen, V. Subbiah, V. Moreno, V. Gambardella, M. Vieito, O. Saavedra, S. Cousin, P. A. Cassier, R. Lobbardi, O. Schönborn-Kellenberger, E. Hurh, M. M. Ruisi and E. I. Buchbinder, *J. Clin. Oncol.*, 2023, **41**, TPS3163.
24. M. Konstantinidou and M. R. Arkin, *Cell Chem. Biol.*, 2024, **31**, 1064–1088.
25. J. M. Sasso, R. Tenchov, D. Wang, L. S. Johnson, X. Wang and Q. A. Zhou, *Biochemistry*, 2023, **62**, 601–623.
26. H. Rui, K. S. Ashton, J. Min, C. Wang and P. R. Potts, *RSC Chem. Biol.*, 2023, **4**, 192–215.
27. X. Tan, L. I. Calderon-Villalobos, M. Sharon, C. Zheng, C. V. Robinson, M. Estelle and N. Zheng, *Nature*, 2007, **446**, 640–645.
28. B. Thines, L. Katsir, M. Melotto, Y. Niu, A. Mandaokar, G. Liu, K. Nomura, S. Y. He, G. A. Howe and J. Browse, *Nature*, 2007, **448**, 661–665.
29. A. Chini, S. Fonseca, G. Fernandez, B. Adie, J. M. Chico, O. Lorenzo, G. Garcia-Casado, I. Lopez-Vidriero, F. M. Lozano, M. R. Ponce, J. L. Micol and R. Solano, *Nature*, 2007, **448**, 666–671.
30. W. Rehman, L. M. Arfons and H. M. Lazarus, *Ther. Adv. Hematol.*, 2011, **2**, 291–308.
31. A. K. Gandhi, J. Kang, C. G. Havens, T. Conklin, Y. Ning, L. Wu, T. Ito, H. Ando, M. F. Waldman, A. Thakurta, A. Klippel, H. Handa, T. O. Daniel, P. H. Schafer and R. Chopra, *Br. J. Haematol.*, 2014, **164**, 811–821.
32. J. Kronke, N. D. Udeshi, A. Narla, P. Grauman, S. N. Hurst, M. McConkey, T. Svinkina, D. Heckl, E. Comer, X. Li, C. Ciarlo, E. Hartman, N. Munshi, M. Schenone, S. L. Schreiber, S. A. Carr and B. L. Ebert, *Science*, 2014, **343**, 301–305.
33. E. R. Watson, S. Novick, M. E. Matyskiela, P. P. Chamberlain, A. H. de la Peña, J. Zhu, E. Tran, P. R. Griffin, I. E. Wertz and G. C. Lander, *Science*, 2022, **378**, 549–553.
34. S. Ichikawa, H. A. Flaxman, W. Xu, N. Vallavoju, H. C. Lloyd, B. Wang, D. Shen, M. R. Pratt and C. M. Woo, *Nature*, 2022, **610**, 775–782.
35. Q. Chen, C. Liu, W. Wang, X. Meng, X. Cheng, X. Li, L. Cai, L. Luo, X. He, H. Qu, J. Luo, H. Wei, S. Gao, G. Liu, J. Wan, D. I. Israel, J. Li and D. Dou, *ACS Chem. Biol.*, 2023, **18**, 25–33.
36. G. W. Collie, M. A. Clark, A. D. Keefe, A. Madin, J. A. Read, E. L. Rivers and Y. Zhang, *J. Med. Chem.*, 2024, **67**, 864–884.
37. H. Richter, A. L. Satz, M. Bedoucha, B. Buettelmann, A. C. Petersen, A. Harmeier, R. Hermosilla, R. Hochstrasser, D. Burger, B. Gsell, R. Gasser, S. Huber, M. N. Hug, B. Kocer, B. Kuhn, M. Ritter, M. G. Rudolph, F. Weibel, J. Molina-David, J. J. Kim, J. V. Santos, M. Stihle,

G. J. Georges, R. D. Bonfil, R. Fridman, S. Uhles, S. Moll, C. Faul, A. Fornoni and M. Prunotto, *ACS Chem. Biol.*, 2019, **14**, 37–49.
38. J. Jang, C. To, D. J. H. De Clercq, E. Park, C. M. Ponthier, B. H. Shin, M. Mushajiang, R. P. Nowak, E. S. Fischer, M. J. Eck, P. A. Jänne and N. S. Gray, *Angew. Chem., Int. Ed.*, 2020, **59**, 14481–14489.
39. R. M. Franzini and C. Randolph, *J. Med. Chem.*, 2016, **59**, 6629–6644.
40. R. Pikalyova, Y. Zabolotna, D. M. Volochnyuk, D. Horvath, G. Marcou and A. Varnek, *Mol. Inf.*, 2022, **41**, e2100289.
41. A. A. Peterson and D. R. Liu, *Nat. Rev. Drug Discovery*, 2023, **22**, 699–722.
42. A. L. Satz, *ACS Med. Chem. Lett.*, 2018, **9**, 408–410.
43. R. I. Troup, C. Fallan and M. G. J. Baud, *Explor. Target. Antitumor Ther.*, 2020, **1**, 273–312.
44. J. S. Disch, J. M. Duffy, E. C. Y. Lee, D. Gikunju, B. Chan, B. Levin, M. I. Monteiro, S. A. Talcott, A. C. Lau, F. Zhou, A. Kozhushnyan, N. E. Westlund, P. B. Mullins, Y. Yu, M. von Rechenberg, J. Zhang, Y. A. Arnautova, Y. Liu, Y. Zhang, A. J. McRiner, A. D. Keefe, A. Kohlmann, M. A. Clark, J. W. Cuozzo, C. Huguet and S. Arora, *J. Med. Chem.*, 2021, **64**, 5049–5066.
45. G. S. Parker, S. Fish, G. Leriche, S. Bailey, J. I. Toth, K. Chng, A. Jamboricic, E. Daniele, S. Chien, S. D. Brown, E. Green, M. Hocker, E. Imler, E. A. Kallel, P. A. Thompson, P. Yang, Y. Zhang and K. Vijayan, AACR, 2021.
46. W. G. Cochrane, M. L. Malone, V. Q. Dang, V. Cavett, A. L. Satz and B. M. Paegel, *ACS Comb. Sci.*, 2019, **21**, 425–435.
47. A. E. Fernandez-Montalvan, M. Berger, B. Kuropka, S. J. Koo, V. Badock, J. Weiske, V. Puetter, S. J. Holton, D. Stockigt, A. Ter Laak, P. A. Centrella, M. A. Clark, C. E. Dumelin, E. A. Sigel, H. H. Soutter, D. M. Troast, Y. Zhang, J. W. Cuozzo, A. D. Keefe, D. Roche, V. Rodeschini, A. Chaikuad, L. Diaz-Saez, J. M. Bennett, O. Fedorov, K. V. M. Huber, J. Hubner, H. Weinmann, I. V. Hartung and M. Gorjanacz, *ACS Chem. Biol.*, 2017, **12**, 2730–2736.
48. bioRender: biorender.com, (accessed June 2024).
49. M. Schroder, M. Renatus, X. Liang, F. Meili, T. Zoller, S. Ferrand, F. Gauter, X. Li, F. Sigoillot, S. Gleim, T. M. Stachyra, J. R. Thomas, D. Begue, M. Khoshouei, P. Lefeuvre, R. Andraos-Rey, B. Chung, R. Ma, B. Pinch, A. Hofmann, M. Schirle, N. Schmiedeberg, P. Imbach, D. Gorses, K. Calkins, B. Bauer-Probst, M. Maschlej, M. Niederst, R. Maher, M. Henault, J. Alford, E. Ahrne, L. Tordella, G. Hollingworth, N. H. Thoma, A. Vulpetti, T. Radimerski, P. Holzer, S. Carbonneau and C. R. Thoma, *Nat. Commun.*, 2024, **15**, 275.
50. A. Hanzl, R. Casement, H. Imrichova, S. J. Hughes, E. Barone, A. Testa, S. Bauer, J. Wright, M. Brand, A. Ciulli and G. E. Winter, *Nat. Chem. Biol.*, 2023, **19**, 323–333.
51. S. W. Kimani, S. Perveen, M. Szewezyk, H. Zeng, A. Dong, F. Li, P. Ghiabi, Y. Li, I. Chau, C. H. Arrowsmith, D. Barsyte-Lovejoy, V. Santhakumar, M. Vedadi and L. Halabelian, *Commun. Biol.*, 2023, **6**, 1272.

52. R. C. Augustin, R. Bao and J. J. Luke, *J. Immunother. Cancer*, 2023, **11**, e006007.
53. H. Cong, L. Xu, Y. Wu, Z. Qu, T. Bian, W. Zhang, C. Xing and C. Zhuang, *J. Med. Chem.*, 2019, **62**, 5750–5772.
54. S. Duan and M. Pagano, *Cell Chem. Biol.*, 2021, **28**, 918–933.
55. L. J. Crawford, D. C. Campbell, J. J. Morgan, M. A. Lawson, J. M. Down, D. Chauhan, R. M. McAvera, T. C. Morris, C. Hamilton, A. Krishnan, K. Rajalingam, A. D. Chantry and A. E. Irvine, *Oncogene*, 2020, **39**, 5001–5014.
56. Y. Jeong, A. R. Oh, Y. H. Jung, H. Gi, Y. U. Kim and K. Kim, *Exp. Mol. Med.*, 2023, **55**, 2097–2104.
57. J. J. N. Veerman, Y. B. Bruseker, E. Damen, E. H. Heijne, W. van Bruggen, K. F. W. Hekking, R. Winkel, C. D. Hupp, A. D. Keefe, J. Liu, H. A. Thomson, Y. Zhang, J. W. Cuozzo, A. J. McRiner, M. J. Mulvihill, P. van Rijnsbergen, B. Zech, L. M. Renzetti, L. Babiss and G. Muller, *ACS Med. Chem. Lett.*, 2021, **12**, 555–562.
58. C. K. Chana, P. Maisonneuve, G. Posternak, N. G. A. Grinberg, J. Poirson, S. M. Ona, D. F. Ceccarelli, P. Mader, D. J. St-Cyr, V. Pau, I. Kurinov, X. Tang, D. Deng, W. Cui, W. Su, L. Kuai, R. Soll, M. Tyers, H. L. Rost, R. A. Batey, M. Taipale, A. C. Gingras and F. Sicheri, *J. Med. Chem.*, 2022, **65**, 12725–12746.
59. A. S. M. Li, S. Kimani, B. Wilson, M. Noureldin, H. Gonzalez-Alvarez, A. Mamai, L. Hoffer, J. P. Guilinger, Y. Zhang, M. von Rechenberg, J. S. Disch, C. J. Mulhern, B. L. Slakman, J. W. Cuozzo, A. Dong, G. Poda, M. Mohammed, P. Saraon, M. Mittal, P. Modh, V. Rathod, B. Patel, S. Ackloo, V. Santhakumar, M. M. Szewczyk, D. Barsyte-Lovejoy, C. H. Arrowsmith, R. Marcellus, M. A. Guie, A. D. Keefe, P. J. Brown, L. Halabelian, R. Al-Awar and M. Vedadi, *J. Med. Chem.*, 2023, **66**, 5041–5060.
60. J. Lee and P. Zhou, *Mol. Cell*, 2007, **26**, 775–780.
61. L. Tan, E. Ehrlich and X. F. Yu, *J. Virol.*, 2007, **81**, 10822–10830.
62. A. Vulpetti, P. Holzer, N. Schmiedeberg, P. Imbach-Weese, C. Pissot-Soldermann, G. J. Hollingworth, T. Radimerski, C. R. Thoma, T. M. Stachyra, M. Wojtynek, M. Maschlej, S. Chau, A. Schuffenhauer, C. Fernandez, M. Schroder and M. Renatus, *ACS Med. Chem. Lett.*, 2023, **14**, 949–954.
63. K. McCloskey, E. A. Sigel, S. Kearnes, L. Xue, X. Tian, D. Moccia, D. Gikunju, S. Bazzaz, B. Chan, M. A. Clark, J. W. Cuozzo, M. A. Guie, J. P. Guilinger, C. Huguet, C. D. Hupp, A. D. Keefe, C. J. Mulhern, Y. Zhang and P. Riley, *J. Med. Chem.*, 2020, **63**, 8857–8866.
64. S. Ackloo, F. Li, M. Szewczyk, A. Seitova, P. Loppnau, H. Zeng, J. Xu, S. Ahmad, Y. A. Arnautova, A. Baghaie, S. Beldar, A. Bolotokova, P. A. Centrella, I. Chau, M. A. Clark, J. W. Cuozzo, S. Dehghani-Tafti, J. S. Disch, A. Dong, A. Dumas, J. A. Feng, P. Ghiabi, E. Gibson, J. Gilmer, B. Goldman, S. R. Green, M.-A. Guié, J. P. Guilinger, N. Harms, O. Herasymenko, S. Houliston, A. Hutchinson, S. Kearnes, A. D. Keefe,

S. W. Kimani, T. Kramer, M. Kutera, H. A. Kwak, C. Lento, Y. Li, J. Liu, J. Loup, R. A. Machado, C. J. Mulhern, S. Perveen, G. L. Righetto, P. Riley, S. Shrestha, E. A. Sigel, M. Silva, M. D. Sintchak, B. L. Slakman, R. D. Taylor, J. Thompson, W. Torng, C. Underkoffler, M. v Rechenberg, I. Watson, D. J. Wilson, E. Wolf, M. Yadav, A. K. Yazdi, J. Zhang, Y. Zhang, V. Santhakumar, A. M. Edwards, D. Barsyte-Lovejoy, M. Schapira, P. J. Brown, L. Halabelian and C. H. Arrowsmith, bioRxiv, 2024, 2024.2003.2003.583197.

65. C. Chaudhry, *J. Med. Chem.*, 2023, **66**, 6239–6250.
66. H. Kiely-Collins, G. E. Winter and G. J. L. Bernardes, *Cell Chem. Biol.*, 2021, **28**, 952–968.
67. A. I. Chan, L. M. McGregor, T. Jain and D. R. Liu, *J. Am. Chem. Soc.*, 2017, **139**, 10192–10195.
68. J. P. Guilinger, A. Archna, M. Augustin, A. Bergmann, P. A. Centrella, M. A. Clark, J. W. Cuozzo, M. Dather, M. A. Guie, S. Habeshian, R. Kiefersauer, S. Krapp, A. Lammens, L. Lercher, J. Liu, Y. Liu, K. Maskos, M. Mrosek, K. Pflugler, M. Siegert, H. A. Thomson, X. Tian, Y. Zhang, D. L. Konz Makino and A. D. Keefe, *Bioorg. Med. Chem.*, 2021, **42**, 116223.
69. Z. Zhu, L. C. Grady, Y. Ding, K. E. Lind, C. P. Davie, C. B. Phelps and G. Evindar, *SLAS Discovery*, 2019, **24**, 169–174.
70. R. Ge, Z. Shen, J. Yin, W. Chen, Q. Zhang, Y. An, D. Tang, A. L. Satz, W. Su and L. Kuai, *SLAS Discovery*, 2022, **27**, 79–85.
71. D. Neri and R. A. Lerner, *Annu. Rev. Biochem.*, 2018, **87**, 479–502.
72. L. Mannocci, M. Leimbacher, M. Wichert, J. Scheuermann and D. Neri, *Chem. Commun.*, 2011, **47**, 12747–12753.
73. R. E. Kleiner, C. E. Dumelin and D. R. Liu, *Chem. Soc. Rev.*, 2011, **40**, 5707–5717.
74. R. A. Goodnow Jr., C. E. Dumelin and A. D. Keefe, *Nat. Rev. Drug Discovery*, 2017, **16**, 131–147.
75. M. Song and G. T. Hwang, *J. Med. Chem.*, 2020, **63**, 6578–6599.
76. D. Madsen, C. Azevedo, I. Micco, L. K. Petersen and N. J. V. Hansen, *Prog. Med. Chem.*, 2020, **59**, 181–249.
77. A. Gironda-Martinez, E. J. Donckele, F. Samain and D. Neri, *ACS Pharmacol. Transl. Sci.*, 2021, **4**, 1265–1279.
78. R. Nussinov, M. Zhang, R. Maloney, Y. Liu, C. J. Tsai and H. Jang, *J. Mol. Biol.*, 2022, **434**, 167569.
79. G. M. Burslem, A. R. Schultz, D. P. Bondeson, C. A. Eide, S. L. Savage Stevens, B. J. Druker and C. M. Crews, *Cancer Res.*, 2019, **79**, 4744–4753.
80. A. Mares, A. H. Miah, I. E. D. Smith, M. Rackham, A. R. Thawani, J. Cryan, P. A. Haile, B. J. Votta, A. M. Beal, C. Capriotti, M. A. Reilly, D. T. Fisher, N. Zinn, M. Bantscheff, T. T. MacDonald, A. Vossenkamper, P. Dace, I. Churcher, A. B. Benowitz, G. Watt, J. Denyer, P. Scott-Stevens and J. D. Harling, *Commun. Biol.*, 2020, **3**, 140.
81. R. P. Wurz, K. Dellamaggiore, H. Dou, N. Javier, M. C. Lo, J. D. McCarter, D. Mohl, C. Sastri, J. R. Lipford and V. J. Cee, *J. Med. Chem.*, 2018, **61**, 453–461.

82. A. K. Charnley, M. A. Convery, A. Lakdawala Shah, E. Jones, P. Hardwicke, A. Bridges, M. Ouellette, R. Totoritis, B. Schwartz, B. W. King, D. D. Wisnoski, J. Kang, P. M. Eidam, B. J. Votta, P. J. Gough, R. W. Marquis, J. Bertin and L. Casillas, *Bioorg. Med. Chem.*, 2015, **23**, 7000–7006.
83. P. A. Haile, B. J. Votta, R. W. Marquis, M. J. Bury, J. F. Mehlmann, R. Singhaus Jr., A. K. Charnley, A. S. Lakdawala, M. A. Convery, D. B. Lipshutz, B. M. Desai, B. Swift, C. A. Capriotti, S. B. Berger, M. K. Mahajan, M. A. Reilly, E. J. Rivera, H. H. Sun, R. Nagilla, A. M. Beal, J. N. Finger, M. N. Cook, B. W. King, M. T. Ouellette, R. D. Totoritis, M. Pierdomenico, A. Negroni, L. Stronati, S. Cucchiara, B. Ziolkowski, A. Vossenkamper, T. T. MacDonald, P. J. Gough, J. Bertin and L. N. Casillas, *J. Med. Chem.*, 2016, **59**, 4867–4880.
84. M. Leveridge, C. W. Chung, J. W. Gross, C. B. Phelps and D. Green, *SLAS Discovery*, 2018, **23**, 881–897.
85. D. P. Bondeson, A. Mares, I. E. Smith, E. Ko, S. Campos, A. H. Miah, K. E. Mulholland, N. Routly, D. L. Buckley, J. L. Gustafson, N. Zinn, P. Grandi, S. Shimamura, G. Bergamini, M. Faelth-Savitski, M. Bantscheff, C. Cox, D. A. Gordon, R. R. Willard, J. J. Flanagan, L. N. Casillas, B. J. Votta, W. den Besten, K. Famm, L. Kruidenier, P. S. Carter, J. D. Harling, I. Churcher and C. M. Crews, *Nat. Chem. Biol.*, 2015, **11**, 611–617.
86. A. H. Miah, I. E. D. Smith, M. Rackham, A. Mares, A. R. Thawani, R. Nagilla, P. A. Haile, B. J. Votta, L. J. Gordon, G. Watt, J. Denyer, D. T. Fisher, P. Dace, P. Giffen, A. Goncalves, I. Churcher, P. Scott-Stevens and J. D. Harling, *J. Med. Chem.*, 2021, **64**, 12978–13003.
87. M. S. Gadd, A. Testa, X. Lucas, K. H. Chan, W. Chen, D. J. Lamont, M. Zengerle and A. Ciulli, *Nat. Chem. Biol.*, 2017, **13**, 514–521.
88. J. W. Mason, Y. T. Chow, L. Hudson, A. Tutter, G. Michaud, M. V. Westphal, W. Shu, X. Ma, Z. Y. Tan, C. W. Coley, P. A. Clemons, S. Bonazzi, F. Berst, K. Briner, S. Liu, F. J. Zecri and S. L. Schreiber, *Nat. Chem. Biol.*, 2024, **20**, 170–179.
89. S. Liu, B. Tong, J. W. Mason, J. M. Ostrem, A. Tutter, B. K. Hua, S. A. Tang, S. Bonazzi, K. Briner, F. Berst, F. J. Zecri and S. L. Schreiber, *J. Am. Chem. Soc.*, 2023, **145**, 23281–23291.
90. X. Liu, C. Huang, X. Wu, L. Meng, X. Lu and X. Li, ChemRxiv, 2024.
91. E. F. Douglass Jr., C. J. Miller, G. Sparer, H. Shapiro and D. A. Spiegel, *J. Am. Chem. Soc.*, 2013, **135**, 6092–6099.
92. P. Cai, B. Sauter, S. Zhanybekova and D. Gillingham, ChemRxiv, 2024.
93. N. Bai, K. M. Riching, A. Makaju, H. Wu, T. M. Acker, S. C. Ou, Y. Zhang, X. Shen, D. N. Bulloch, H. Rui, B. W. Gibson, D. L. Daniels, M. Urh, B. M. Rock and S. C. Humphreys, *J. Biol. Chem.*, 2022, **298**, 101653.
94. S. L. Fisher and A. J. Phillips, *Curr. Opin. Chem. Biol.*, 2018, **44**, 47–55.
95. S. Bonazzi, E. d'Hennezel, R. E. J. Beckwith, L. Xu, A. Fazal, A. Magracheva, R. Ramesh, A. Cernijenko, B. Antonakos, H. C. Bhang, R. G. Caro, J. S. Cobb, E. Ornelas, X. Ma, C. A. Wartchow, M. C. Clifton,

R. R. Forseth, B. H. Fortnam, H. Lu, A. Csibi, J. Tullai, S. Carbonneau, N. M. Thomsen, J. Larrow, B. Chie-Leon, D. Hainzl, Y. Gu, D. Lu, M. J. Meyer, D. Alexander, J. Kinyamu-Akunda, C. A. Sabatos-Peyton, N. A. Dales, F. J. Zecri, R. K. Jain, J. Shulok, Y. K. Wang, K. Briner, J. A. Porter, J. A. Tallarico, J. A. Engelman, G. Dranoff, J. E. Bradner, M. Visser and J. M. Solomon, *Cell Chem. Biol.*, 2023, **30**, 235–247. e212.

96. X. Han and Y. Sun, *Cell Rep.*, 2022, **3**, 101062.
97. J. P. Daguer, C. Zambaldo, D. Abegg, S. Barluenga, C. Tallant, S. Muller, A. Adibekian and N. Winssinger, *Angew. Chem., Int. Ed.*, 2015, **54**, 6057–6061.
98. C. Zambaldo, J. P. Daguer, J. Saarbach, S. Barluenga and N. Winssinger, *MedChemComm*, 2016, **7**, 1340–1351.
99. G. Zimmermann, U. Rieder, D. Bajic, S. Vanetti, A. Chaikuad, S. Knapp, J. Scheuermann, M. Mattarella and D. Neri, *Chemistry*, 2017, **23**, 8152–8155.
100. C. E. Hendrick, J. R. Jorgensen, C. Chaudhry, I. I. Strambeanu, J. F. Brazeau, J. Schiffer, Z. Shi, J. D. Venable and S. E. Wolkenberg, *ACS Med. Chem. Lett.*, 2022, **13**, 1182–1190.
101. R. Stevens, E. Bendito-Moll, D. J. Battersby, A. H. Miah, N. Wellaway, R. P. Law, P. Stacey, D. Klimaszewska, J. M. Macina, G. A. Burley and J. D. Harling, *J. Med. Chem.*, 2023, **66**, 15437–15452.
102. Z. Wang, S. Shaabani, X. Gao, Y. L. D. Ng, V. Sapozhnikova, P. Mertins, J. Kronke and A. Domling, *Nat. Commun.*, 2023, **14**, 8437.
103. Z. Kozicka, D. J. Suchyta, V. Focht, G. Kempf, G. Petzold, M. Jentzsch, C. Zou, C. Di Genua, K. A. Donovan, S. Coomar, M. Cigler, C. Mayor-Ruiz, J. L. Schmid-Burgk, D. Haussinger, G. E. Winter, E. S. Fischer, M. Slabicki, D. Gillingham, B. L. Ebert and N. H. Thoma, *Nat. Chem. Biol.*, 2024, **20**, 93–102.
104. C. Mayor-Ruiz, S. Bauer, M. Brand, Z. Kozicka, M. Siklos, H. Imrichova, I. H. Kaltheuner, E. Hahn, K. Seiler, A. Koren, G. Petzold, M. Fellner, C. Bock, A. C. Muller, J. Zuber, M. Geyer, N. H. Thoma, S. Kubicek and G. E. Winter, *Nat. Chem. Biol.*, 2020, **16**, 1199–1207.
105. L. Jin and S. C. Harrison, *Proc. Natl. Acad. Sci. U. S. A.*, 2002, **99**, 13522–13526.
106. A. Marchand, S. Buckley, A. Schneuing, M. Pacesa, P. Gainza, E. Elizarova, R. M. Neeser, P.-W. Lee, L. Reymond, M. Elia, L. Scheller, S. Georgeon, J. Schmidt, P. Schwaller, S. J. Maerkl, M. Bronstein and B. E. Correia, bioRxiv, 2024, 2024.2003.2025.585721.
107. Y. Dong, T. Ma, T. Xu, Z. Feng, Y. Li, L. Song, X. Yao, C. R. Ashby and G.-F. Hao, *Acta Pharm. Sin. B*, 2024, **14**, 4266–4295.
108. G. Nishiguchi, F. Keramatnia, J. Min, Y. Chang, B. Jonchere, S. Das, M. Actis, J. Price, D. Chepyala, B. Young, K. McGowan, P. J. Slavish, A. Mayasundari, J. A. Jarusiewicz, L. Yang, Y. Li, X. Fu, S. H. Garrett, J. B. Papizan, K. Kodali, J. Peng, S. M. Pruett Miller, M. F. Roussel, C. Mullighan, M. Fischer and Z. Rankovic, *J. Med. Chem.*, 2021, **64**, 7296–7311.

109. K. L. Thomas, H. Bouguenina, D. S. J. Miller, F. J. Sialana, T. G. Hayhow, J. S. Choudhary, O. W. Rossanese and B. R. Bellenie, *ACS Chem. Biol.*, 2024, **19**, 173–184.
110. B. Z. Stanton, E. J. Chory and G. R. Crabtree, *Science*, 2018, **359**, eaao5902.
111. S. Singh, W. Tian, Z. C. Severance, S. K. Chaudhary, V. Anokhina, B. Mondal, R. Pergu, P. Singh, U. Dhawa, S. Singha and A. Choudhary, *Chem. Soc. Rev.*, 2023, **52**, 5485–5515.
112. X. Liu and A. Ciulli, *ACS Cent. Sci.*, 2023, **9**, 1269–1284.
113. R. Amirian, M. Azadi Badrbani, Z. Izadi, H. Samadian, G. Bahrami, S. Sarvari, S. Abdolmaleki, S. M. Nabavi, H. Derakhshankhah and M. Jaymand, *Eur. J. Med. Chem.*, 2023, **260**, 115765.
114. C. Bashore, S. Prakash, M. C. Johnson, R. J. Conrad, I. A. Kekessie, S. J. Scales, N. Ishisoko, T. Kleinheinz, P. S. Liu, N. Popovych, A. T. Wecksler, L. Zhou, C. Tam, I. Zilberleyb, R. Srinivasan, R. A. Blake, A. Song, S. T. Staben, Y. Zhang, D. Arnott, W. J. Fairbrother, S. A. Foster, I. E. Wertz, C. Ciferri and E. C. Dueber, *Nat. Chem. Biol.*, 2023, **19**, 55–63.
115. E. M. H. Ali, C. A. Loy and D. J. Trader, bioRxiv, 2024, 2024.2001.2020.576376.
116. G. Ahn, S. M. Banik, C. L. Miller, N. M. Riley, J. R. Cochran and C. R. Bertozzi, *Nat. Chem. Biol.*, 2021, **17**, 937–946.
117. C. H. Ji, H. Y. Kim, M. J. Lee, A. J. Heo, D. Y. Park, S. Lim, S. Shin, S. Ganipisetti, W. S. Yang, C. A. Jung, K. Y. Kim, E. H. Jeong, S. H. Park, S. Bin Kim, S. J. Lee, J. E. Na, J. I. Kang, H. M. Chi, H. T. Kim, Y. K. Kim, B. Y. Kim and Y. T. Kwon, *Nat. Commun.*, 2022, **13**, 904.

Subject Index